Reinventing
Environmental Regulation

Reinventing Environmental Regulation

Lessons from Project XL

Alfred A. Marcus
Donald A. Geffen
Ken Sexton

RESOURCES FOR THE FUTURE ≡ WASHINGTON, DC

Printed in the United States of America

An RFF Press book
Published by Resources for the Future
1616 P Street, NW, Washington, DC 20036–1400

A publication of Resources for the Future (www.rff.org).

Library of Congress Cataloging-in-Publication Data
Marcus, Alfred Allen, 1950–
    Reinventing environmental regulation : lessons from Project XL / Alfred A. Marcus, Donald A. Geffen, and Ken Sexton.
        p.   cm.
    Includes bibliographical references and index.
    ISBN 1-891853-08-2 (cloth: alk. paper) — ISBN 1-891853-09-0 (pbk. : alk. paper)
    1. Environmental management—United States.   2. Industries—Environmental aspects—United States.   3. Environmental law—United States.   4. Pollution—Law and legislation—United States.   5. Factory and trade waste—Law and legislation—United States.   I. Geffen, Donald A.   II. Sexton, Ken.   III. Title.
GE310 .M37 2002
363.7'056'0973—dc21   2002010848

f e d c b a

The paper in this book meets the guidelines for permanence and durability of the Committee on Production Guidelines for Book Longevity of the Council on Library Resources.

The text of this book was designed and typeset by Betsy Kulamer in Trump Medieval and ITC Franklin Gothic. It was copyedited by Alfred F. Imhoff. The cover was designed by Naylor Design.

ISBN 1–891853–08–2 (cloth)   ISBN 1–891853–09–0 (paper)

About Resources for the Future and RFF Press

Resources for the Future (RFF) improves environmental and natural resource policymaking worldwide through independent social science research of the highest caliber.

Founded in 1952, RFF pioneered the application of economics as a tool to develop more effective policy about the use and conservation of natural resources. Its scholars continue to employ social science methods to analyze critical issues concerning pollution control, energy policy, land and water use, hazardous waste, climate change, biodiversity, and the environmental challenges of developing countries.

RFF Press supports the mission of RFF by publishing book-length works that present a broad range of approaches to the study of natural resources and the environment. Its authors and editors include RFF staff, researchers from the larger academic and policy communities, and journalists. Audiences for RFF publications include all of the participants in the policymaking process—scholars, the media, advocacy groups, nongovernmental organizations, professionals in business and government, and the general public.

Contents

Whether or not anything can be known
can be settled not by arguing
but by trying.

———

Francis Bacon

1

Environmental Policy in Transition

During the 1990s, environmental policy in the United States began a transition that continues to this day. Long-standing assumptions underlying the conventional command-and-control regulatory system have come under increasing scrutiny. Diverse individuals and organizations—including business leaders, regulators, elected officials, policy analysts, and environmentalists—want to find more effective, efficient, and equitable ways to protect environmental quality and public health (Fischbeck and Farrow 2001; Coglianese and Nash 2001; Sexton et al. 1999; Chertow and Esty 1997; Howard 1994; John 1994). In the midst of this national policy debate, the U.S. Environmental Protection Agency (EPA) initiated a series of regulatory reform experiments to test whether governments and businesses could work together to adopt innovative approaches to achieve improved environmental results at lower cost. This book tells the story of how one of these experiments laid bare intrinsic problems that often become roadblocks to cooperative solutions.

Our story begins in 1995 and involves two major protagonists: the 3M Company, widely recognized as environmentally progressive and socially responsible, and EPA, the federal focal point for environmental regulation. EPA was just starting to implement Project XL (for eXcellence and Leadership), which was one of President Bill Clinton's and Vice President Al Gore's flagship projects aimed at "reinventing environmental regulations" (Clinton and Gore 1995, i) to make pollution control "cleaner, cheaper, and smarter" (Clinton and Gore 1995, 5). Under XL, a company or other regulated entity could propose a demonstration project that would replace existing regulatory requirements with more flexible ones, provided that the company or other regulated entity achieved better environmental results than would have been expected under the old requirements.

In response to EPA's solicitation for XL proposals, 3M requested permission to test a creative, unconventional approach to meet air pollution control requirements at its Hutchinson, Minnesota, plant, which was a state-of-the-art facility manufacturing audio, video, and adhesive tape products. The proposal called for setting fixed plantwide limits on the amount of air pollution that could be emitted but allowed 3M to make changes at the plant without obtaining permits that would impose costly delays. If 3M were to be granted this operating flexibility, its managers claimed that they would be better able to explore innovative environmental management techniques while keeping emissions well below existing regulatory limits.

The prospects for the success of 3M's proposal seemed bright. But in reality the negotiations were contentious and protracted, and they eventually broke down. We chronicle the collapse of these negotiations and discuss what happened in the context of XL projects that EPA carried out with other companies—Intel Corporation, Weyerhaeuser Company, and Merck & Company. We explore the reasons for the 3M stalemate and compare it to the other cases. Based on our analysis, we propose modifying EPA's approach to reinvention to revitalize XL-like experiments and increase their ability to yield fruitful results.

Finding Common Ground

Shortly after Christine Todd Whitman became EPA administrator in early 2001, she stated that finding "common ground" was necessary to achieve the agency's goals of preserving, protecting, and improving the environment while strengthening the economy.[1] She cited Project XL as a prime example of EPA's efforts to achieve better environmental protection at lower cost by negotiating site-specific environmental agreements.[2] Whitman went on to describe XL as an "innovative and incentives-based program" that provided "regulatory flexibility to companies that agree to meet ... environmental goals that exceed regulatory standards." She said it was a "model" of how EPA should work—building partnerships with stakeholders, focusing on results, and moving away from conventional command-and-control approaches to more cooperative partnerships.

Project XL had started six years earlier on March 16, 1995, under Carol Browner, Whitman's predecessor at EPA. XL was one of the most ambitious and potentially consequential U.S. experiments seeking common ground in environmental policymaking. It was one of 25 actions taken by the Clinton administration to explore innovative alternatives to existing regulations. It emphasized partnerships among businesses, governments, and other stakeholders as a way to achieve enhanced environmental protection while maintaining economic prosperity. Moreover, it was a centerpiece of the administration's effort "to develop innovative alternatives to the current regulatory system" (Clinton and Gore 1995). To implement it, EPA would "enter into partnerships with businesses, environmentalists, states and communities to test alternative environmental protection strategies" in "a coordinated series of projects." "The knowledge gained from such bold experimentation," according to the Clinton administration, would "lay the groundwork for developing a new environmental management system for the 21st century."

The problem was how to put this admirable idea into practice, especially when it depended on getting long-standing antagonists to work together collaboratively for environmental benefit. In this book, our goal is to examine factors that affected the search for common ground in negotiating experimental environmental accords and to recommend ways to improve the chances that accords of this nature will be reached and experiments carried out. The focus initially is on the efforts of 3M, EPA, the Minnesota Pollution Control Agency (MPCA), and MPCA's stakeholder group, the Pilot Project Committee (PPC), to achieve a mutually amenable agreement. As both participants and observers, we were involved in the negotiations and had almost unlimited access to the key participants and to project-related information. We helped to create and served as members of the stakeholder group that advised MPCA on the project. At the same time, we formed a research team with support from an EPA grant and in this capacity monitored and evaluated the project (PPC, notes from meetings of the Research Committee, 1996–97; Marcus, Geffen, Sexton, and Wiessner 1998). In these dual roles, we attended key project meetings, carried out interviews with MPCA, 3M, and EPA staff, collected documents and other archival evidence, and conducted a survey.[3]

In this book, we use the stalemate that developed between 3M and EPA to illustrate the real-life obstacles that must be overcome when institutions that often have been at odds over environmental issues try to set aside their differences to achieve cooperative solutions. The Intel, Merck, and Weyerhaeuser projects were less challenging than 3M's, but reaching an agreement for each took longer and was more difficult than the parties anticipated (Coglianese 1997; Stewart 2001). We examine how the participants in these projects found common ground, whereas the 3M pilot ended in deadlock.

Toward a New Paradigm

Project XL is but one of a new generation of cooperative and collaborative approaches to environmental policy that have emerged in the past decade.

Most analysts have acknowledged that the United States has made substantial progress in reducing pollution from the largest and most obvious sources (Sexton et al. 1999). Nonetheless, many analysts also have maintained that the existing environmental management system should be overhauled to meet the complex challenges of the future (Breyer 1993; John 1994; NAPA 1995, 1997; Sexton et al. 1999).[4] As Phillip Howard (1994, 11) noted in *The Death of Common Sense*, "We seem to have achieved the worst of both worlds: a system of regulation that goes too far while it also does too little."

Cooperative approaches exist in the domain between command-and-control regulations, often favored by environmentalists, and market-based incentives, which have long been advocated by economists. These new approaches start from the premise that top-down standards have reached a limit in terms of what they can accomplish and that the bottom-up market techniques advocated by economists (for instance, buying and selling pollution rights and imposing pollution taxes) are not yet practical.[5] Some see cooperative approaches as a replacement for conventional regulation, others as an alternative to conventional regulation, and still others as a reinforcement of conventional regulation. These approaches require that businesses and governments cooperate with each other as well as with nongovernmental organizations and local communities to reach negotiated solutions to pressing environmental problems (Sexton et al. 1999). The intent is for businesses, governments, and other affected or interested parties to obligate themselves to actions that will simultaneously improve the environment and lower the associated costs for the public and private sectors (Delmas and Terlaak 2000; Hirsch 2001). In the end, these approaches constitute a new paradigm for environmental management (see Table 1-1) that involves a significant shift in conventional policies and practices (Sexton, Murdock, and Marcus 2002).

In concert with these changes, during the past decade the private sector has moved toward greater intrafirm cooperation. In the business world, greater cooperation has emerged between suppliers and customers. Use of joint ventures, alliances, and other partnerships among organizations that had been competitors is also growing (Ring and Van de Ven 1992, 1994; Smith et al. 1995). Parallel processing—the pragmatic sharing and pooling of ideas and perspectives in face-to-face encounters among people from groups traditionally isolated from each other—is quicker and has better outcomes (Schilling and Hill 1998). Unfortunately, businesses have not adopted this cooperative approach when dealing with regulators, nor have government agencies picked up on the idea. Susskind and others (2000, 30) write that "it is ironic that corporations must be trained to behave in this way with regulators ... there is good reason to treat the regulator as a business partner, as important as any other key supplier or customer." Historically, environmental problems in the United States have been managed sequentially; those

Table 1-1. A Comparison of Conventional and Collaborative Paradigms

Conventional Paradigm (Adversarial Relationships)	Collaborative Paradigm (Cooperative Solutions)
Regulators versus industry	Collaborative stakeholder partnerships
Command-and-control strategies	Cooperative, voluntary agreements
End-of-pipe controls	Focus on pollution prevention first
Narrow, media-based statutes	Holistic, multimedia approaches
One-size-fits-all regulations	Place-based environmental decisions
Rigid, prescriptive rules	Flexible, easy-to-adjust rules
Means(process)-based standards	Outcome(results)-based standards
Limited use of market mechanisms	Increased use of market mechanisms
Regulators decide, announce, defend decisions	Stakeholders meaningfully involved

Source: Sexton, Murdock, and Marcus 2002.

involved made separate decisions at distinct points in time. Though the differences between businesses and governments are obvious, the incorporation of parallel procedures into environmental management policies and practices could help to foster less costly solutions tailored to particular problems (Michael 1996). Advocates of this approach believe that the parties can address critical problems that might not be solved by other means (Clayton and Radcliffe 1996; Chertow and Esty 1997; Sexton et. al. 1999).

Cooperative agreements can be more integrated than typical regulatory permits, less dependent on the medium-based approach (air, water, land) that Congress has established, and more involving of citizens (Arthur D. Little 1996; Clayton and Radcliffe 1996; Chertow and Esty 1997; Sexton et. al. 1999). About the practicality of cooperative solutions, however, questions remain. They are but a single tool for solving environmental problems, and the record of past efforts to forge cooperative solutions is mixed.[6]

At the global level, the success of the Montreal Protocols brought worldwide use of chlorofluorocarbons under control. This agreement reversed a seemingly hopeless stalemate, and its outcomes far exceeded expectations. The parties showed great flexibility in creating a high-quality solution to a stubborn environmental problem. However, a similar breakthrough has not been achieved for global warming, despite years of international effort.

On a local level, one form of cooperative problem solving that has met with some success is alternative dispute resolution, which seeks to resolve disputes more expeditiously and with less cost. This technique involves consensus building, joint problem-solving, and negotiation. The parties have direct face-to-face interaction, and they reach decisions by mutual agreement. They jointly explore and resolve their differences with a third-party mediator or facilitator, who generally has no formal authority to impose an outcome.

They can always withdraw and seek resolution through other means. However, alternative dispute resolution has not always been successful.[7]

At the national level, the 1990 Negotiated Rule-Making Act established a statutory basis for the use of regulatory negotiation. Though negotiated rulemaking has been thought to reduce the average time of rulemaking, it was used very infrequently, in less than one-tenth of one percent of 1983–1996 cases (Coglianese 1997). Coglianese (1999) reported that even advocates of negotiated rulemaking claim it is appropriate in very few cases—less than 5%, according to some estimates.

As former EPA Administrator William Ruckelshaus (1996, 3) observed, cooperative approaches are not "panaceas for every environmental problem.... They are extremely difficult to bring off, frustrating to participate in, often lengthy, and expensive for their members, and they can easily fail." Few people deny that conventional command-and-control approaches still will have a place.

O'Leary and others (1999) have made the salient point that environmental disputes are more difficult to negotiate than labor disputes. Environmental disputes do not involve two clearly identified interests of roughly equal stature. The issues are not necessarily concrete, like pay and benefits, and no highly institutionalized process with a long history is routinely used for the resolution of differences. Many different groups are involved—ranging from state and national governments to national and local environmental groups, major corporations, and local citizens. Moreover, the issues are hard to quantify (for instance, irreversible harm to future generations).

O'Leary (1995, 17–18) maintains that there is a paucity of "evidence-based work" on cooperative approaches, the empirical foundations are "weak," and much of what we know is based on "thoughtful speculation." Though cooperative solutions have the potential to break impasses that have prevented the resolution of environmental problems, real-world experience in creating workable agreements is limited (Osborne and Gabler 1993; Bacow and Wheeler 1984; Breyer 1993; Howard 1994; Sexton et al. 1999). Additional analysis is necessary to determine realistically what works and what does not. By comparing the 3M case (see Table 1-2) to the Intel, Weyerhaeuser, and Merck cases, we try to address this shortcoming.

Notes

[1]Her speech was made to the National Environmental Policy Institute on March 8, 2001.

[2]The agency had other initiatives in process (such as Performance Track), and programs of this nature, sometimes quite successfully, were being carried out in a number of states.

Notes continue on page 9

Table 1-2. Timeline for Project XL–Minnesota

Policy Formulation	
National Level	*State Level: Minnesota*
Jan. 1993: Start of Aspen Institute quarterly meetings on environmental issues that last until 1996	
	March 1993: Flexible permit granted to 3M's Saint Paul (Bush Street) facility
June 1993: Creation of Clinton Administration's Council on Sustainable Development	
July 1994: Start of Common Sense Initiative (CSI)	
	Oct. 1994: 3M release of draft of Beyond Compliance Emissions Reduction Act; presentations to various groups
	Jan. 1995: Pollution Prevention (P2) Dialogue discussion of initiating a pilot
March 1995: Announcement of Reinventing Environmental Regulation with Project XL as key component	
May 1995: Project XL *Federal Register* Notice	*May 1995:* P2 Dialogue creation of Pilot Project Committee (PPC) to be involved in XL-like experiments
	June 1995: MPCA's Project XL application to be given the authority to undertake 3–5 pilots
	July 1995: 3M's Project XL application for sites in Minnesota (Hutchinson), Illinois, and California
	Aug. 1995: Lisa Thorvig appointed Minnesota Pollution Control Agency (MPCA) XL Coordinator
Sept. 1995: Environmentalists letter to Gore on XL and CSI	*Sept. 1995:* MPCA meetings seeking additional XL applicants
Oct. 1995: Region 5's draft Memorandum of Understanding (MOU) sent to MPCA	*Oct. 1995:* Thorvig and Dave Ullrich's (EPA Region 5 Deputy Administrator) tour of Hutchinson
Nov. 1995: EPA's response to environmentalists	

Continued on next page

Table 1-2. Timeline for Project XL–Minnesota *(continued)*

Policy Formulation (continued)

National Level	State Level: Minnesota
Nov. 1995: Clinton announcement of first eight XL Project selections, including MPCA and 3M	
Nov. 1995–January 1996: Government shutdown: federal budget stalemate	

Policy Implementation

Minnesota	Minnesota and EPA
Nov. 1995: 3M announcement of major corporate restructuring	
Dec. 1995: Kickoff meeting for MPCA and 3M teams working on Hutchinson agreement	*Dec. 1995:* MPCA's negative response to Region 5's draft MOU
Jan. 1996: 3M "Covenant" (its draft Hutchinson permit) presented to MPCA	
Feb.–March 1996: MPCA's stakeholder group (the PPC) briefed on draft "Covenant"	
March 1996: Passage of Environmental Regulatory Innovations Act by Minnesota State legislature	
March–May 1996: Draft of proposed XL permit and Final Project Agreement (FPA) for Hutchinson completed by MPCA	*May 1996:* MPCA, 3M, and PPC meeting with Region 5 in Chicago to discuss proposed permit and FPA
	May 1996: Proposed permit and FPA put on Public Notice by MPCA
	June 1996: Minnesota stakeholders meet
	July 1996: Response of EPA and Natural Resources Defense Council in Chicago with EPA staff from headquarters and Region 5
	July 1996: EPA, MPCA, 3M, and stakeholder meeting to discuss stalemate
	Aug.–Nov. 1996: MPCA and EPA efforts to restart project
	Dec. 1996: 3M withdrawal from XL and return to regular regulatory system

[3]For a discussion of the kinds of methods we used, see Yin 1989; Pettigrew 1985; Abbott 1990, 1992; Eisenhardt 1989; Van de Ven 1992; and Grazman and Van de Ven 1992.

[4]A National Academy of Public Administration report to Congress (NAPA 1995; as cited by Sexton et. al. 1999, 101) maintained: "To continue to make environmental progress, the nation will have to develop a more rational, less costly strategy for protecting the environment, one that achieves its goals more efficiently, using more creativity and less bureaucracy."

[5]Even market-based approaches require some centralized administration. For instance, under a scheme to buy and sell pollution rights, someone must establish standards, allocate the rights, and maintain the market. Likewise, under a scheme of pollution taxes, someone must determine how high the taxes should be on the basis of the harm caused by the pollution and the extent of emissions into ecosystems of varying sensitivity.

[6]For instance, Marcus and others (1984) found that issues that ranked highest in potential benefits ranked lowest with respect to the feasibility of reaching agreement.

[7]Before Bingham's assessment (Bingham 1986), little systematic evidence had been collected on the benefits and weaknesses of alternative dispute resolution. Bingham examined 132 cases. Three-quarters involved site-specific issues, and one-quarter involved broad policy issues. The site-specific cases typically aimed to resolve controversies about resource use or a toxic waste cleanup, whereas the policy dialogues tried to clarify positions and generate new options about policy issues without immediate change in laws or regulations. Bingham reported that agreement had been reached in a high percentage of the cases she examined—79% of the site-specific cases and 76% of the policy issues—and that 80% of the site-specific cases and 41% of the policy cases were fully implemented. The main reasons for the high rate of success had to do with the substance of the issues involved and the procedures followed. Facilitators pre-screened the situation. Before entering a negotiation, they tried to determine if a positive outcome was feasible.

2

Quid Pro Quo and the Birth of Project XL

This chapter shows that the framework for reaching a cooperative environmental accord can be an obstacle to reaching a workable agreement. If the framework is vague, indefinite, and subject to interpretation, an agreement will be difficult to reach. The framework under Project XL was in the form of a quid pro quo. In exchange for superior environmental performance, a facility seeking a permit would be granted regulatory flexibility. This concept was at the heart of the ideas that went into XL. The Clinton administration introduced it, and the U.S. Environmental Protection Agency (EPA) used it in carrying out the program. Just as we show its importance as the foundation for XL, we also show that it had limitations that hindered efforts to reach cooperative solutions.

Shifts in Public Policy

As discussions of environmental policy in the mid-1990s moved toward experimenting with new approaches, they were marked by political confrontations. A defining event was the 1994 congressional elections, which gave Republicans control of both the U.S. House of Representatives and the U.S. Senate for the first time in 40 years. Given the two major political parties' conflicting philosophies and worldviews, it is not surprising that they disagreed about methods for managing environmental pollution. These differences influenced how they approached questions of environmental management and were at the root of the political debate over environmental policy in the mid-1990s.

The 1994 congressional elections produced a shift in policy as victorious Republicans, especially in the House, sought to implement a regulatory relief and rollback agenda. They attempted to achieve their objectives by pursuing three complementary strategies:

1. rewriting major environmental laws;
2. using riders on appropriations and budget bills to limit implementation and enforcement of existing regulations and, at the same time, cutting EPA's budget; and
3. restructuring the regulatory process itself to slow down regulatory activity.

The first strategy became bogged down in political gridlock, and, except for the Safe Drinking Water Act of 1996, major environmental laws were not rewritten. The second strategy succeeded in shutting down federal regulatory activities for brief periods, but they quickly returned to their normal levels and most legislative riders were vetoed. This strategy backfired when the press and the American public blamed the Republican Congress for playing political games and preventing the federal government from functioning. The third strategy, restructuring the regulatory process, became the focus of political action and debate.

The nature and direction of the environmental policy shift that began in the 104th Congress were epitomized by the Republican manifesto, the Contract with America. In 1994, as part of their election campaign, 367 Republican candidates for the House of Representatives signed the Contract with America, thereby pledging to "roll back government regulations and create jobs." A central feature of the contract was the Job Creation and Wage Enhancement Act, which was introduced at the beginning of the 104th Congress. This bill included provisions to:

- ensure that more scientific and economic analyses were performed;
- increase opportunities for regulated industries to help shape regulatory provisions;
- make certain that only relatively serious risks are regulated;
- require proof that the benefits resulting from regulations exceed the costs of compliance and that proposed regulations represent the most cost-effective option;
- create a regulatory moratorium on the issuance of new regulations until the regulatory reform agenda is enacted;
- establish a regulatory "budget" that places a cap on compliance costs to be imposed on industry;

- change the way federal programs are funded, so that unfunded federal mandates require additional votes by Congress;
- require federal agencies to compensate property owners for losses in property values resulting from environmental regulation; and
- increase procedural protections for those subject to regulatory inspections and enforcement.

Although Republican proposals for change elicited much debate and many legislative initiatives passed one or both chambers, the party made little progress in implementing its agenda. An exception was the Unfunded Mandates Reform Act, which was passed by Congress and signed into law by President Clinton in 1995. This law makes it more difficult for Congress to impose regulatory responsibilities on state and local governments without appropriating adequate resources to do the job. None of the other elements of the Contract with America were enacted into law. Realizing that its legislative agenda was stalled, faced with obvious public antipathy, and fearing political harm, many in the party backed off.

In January 1996, 30 Republican moderates in Congress wrote to Speaker of the House Newt Gingrich to complain that the party had suffered from "missteps in environmental policy" and to ask him to remedy the situation (Sexton and Murdock 1996). The letter warned, "If the party is to resuscitate its reputation in this important area, we cannot be seen as using the budget crisis as an excuse to emasculate environmental protection." Subsequently, in May 1996, House Republicans issued a new vision statement on environmental policy, which was designed to improve the party's environmental image by establishing principles and guidelines for future legislative proposals. Among the principles espoused in the statement were the following (Sexton and Murdock, 1996, 66–67):

- environmental regulations should set common-sense standards without dictating the precise technologies for meeting those standards;
- federal policies should emphasize incentives rather than setting down inflexible laws;
- states and localities should play a greater role in setting and enforcing environmental standards; and
- environmental decisions should be based on consensus and made in consultation with people whose homes, businesses, and communities are directly affected.

The Democrats' Rejoinder

The Democrats, particularly officials in the White House and at EPA, were concerned about the results of the 1994 elections, fearing that the pendulum of public opinion had swung back from overwhelming and unswerving support for environmental regulations to a serious questioning of the trade-offs between environmental benefits and economic costs. These concerns prompted meetings among high-level administration officials to decide how to react.

President Clinton issued a rejoinder to the Republicans in his January 24, 1995, State of the Union Address:

> Do we need more common sense and fairness in our regulations? You bet we do. But we can have common sense and still provide safe drinking water. We can have fairness and still clean up toxic waste dumps. And we ought to do it. (Clinton and Gore, 1995, 1)

In response to the Contract with America, on March 16, the Clinton administration released a manifesto on environmental policy, titled *Reinventing Environmental Regulation*.[1] In this document, Clinton and Gore promised to "oppose those who would undercut protection of public health and the environment under the guise of 'regulatory relief'." Most important, however, the administration admitted that changes were needed and called for "reinventing environmental protection so it will protect more and cost less." The document went on to advocate adoption of commonsense reforms of the existing environmental regulatory system, and it called for new strategies and regulations that were "cleaner, cheaper, and smarter." As Clinton and Gore noted, "We will work with the new Congress whenever possible, but we will not go backwards. Reinvention yes, rollback no" (Clinton and Gore 1995, 5).

Reinventing Regulation

The Clinton administration regulatory reinvention initiatives called for widespread experimentation with new methods that emphasized trust and teamwork between businesses and governments (Sabel 1991; Hosmer 1995; Mayer and Davis 1995; Ruckelshaus 1996; Weber 1998). In *Reinventing Environmental Regulation*, the administration said it was committed to ending adversarial regulation and to ushering in a new era based on flexibility and collaboration. The one-way traffic—from government to business—would have to become two-way (Sparrow 1998). The boundaries between sectors historically at odds would have to become more permeable.

The administration was also direct and candid in its assessment of the limitations of the current regulatory system. New policy tools were needed to encourage technological innovation, lower the costs of regulation, and achieve environmental benefits. Looking ahead, the administration sketched a vision of how environmental regulation would evolve during the next 25 years. Its emphasis was on economic growth and performance, not simple compliance. It wanted to align environmental protection with business strategy and provide local decisionmakers with greater authority.

The administration listed the main principles that would guide the formulation, implementation, and evaluation of its plan for reinventing environmental protection (Clinton and Gore, 1995, 6).

- Protecting public health and the environment are important national goals, and individuals, businesses, and government must take responsibility for the impact of their actions.
- Regulation must be designed to achieve environmental goals in a manner that minimizes costs to individuals, businesses, and other levels of government.
- Environmental regulations must be performance based, providing maximum flexibility in the means of achieving our environmental goals, but requiring accountability for the results.
- Preventing pollution, not just controlling or cleaning it up, is preferred.
- Market incentives should be used to achieve environmental goals, whenever appropriate.

Among the administration's other principles were that "decision making should be collaborative, not adversarial" and that federal and state governments should "work as partners to achieve common environmental goals." The administration described a strategy for environmental regulatory reinvention with 25 high-priority actions to achieve its goals. Among these actions were two tracks. The first involved 18 actions, which included an open market for air pollutant emissions, effluent trading in watersheds, expanded use of risk assessment in local communities, regulatory negotiation and consensus-based rulemaking, paperwork reduction, one-stop emissions reports, consolidated rules, compliance incentives, and self-certification. The second track was designed to develop new and innovative alternatives. It involved partnerships with businesses, environmentalists, states, and communities to test alternative management strategies. The knowledge gained from such "bold experimentation" was supposed to "lay the groundwork for developing a new environmental management system for the 21st century" (Clinton and Gore 1995, 7, 17).

Project XL

The first item listed under the second track of regulatory reinvention initiatives was Project XL. Among the regulatory initiatives the Clinton administration proposed, Clinton singled it out for special mention:

The most notable of these initiatives is Project XL. This program will give ... responsible companies the opportunity to demonstrate *excellence and leadership*. They will be given the flexibility to develop alternative strategies that will replace current regulatory requirements, while producing even greater environmental benefits.... This project is a critical component of the Administration's effort to reinvent regulation. (Clinton and Gore 1995, 5, 14)

According to the Clinton–Gore plan, if a company had a record of good environmental stewardship, then the government would offer to support, on a demonstration basis, company proposals to replace existing regulatory requirements with creative ways of achieving better environmental results than could be expected under existing rules and regulations. To be eligible for consideration under Project XL, a company's proposed pilot or demonstration project had to meet certain conditions. These included the requirement that the alternative strategy produce "environmental performance superior to that which would be achieved by full compliance with current laws and regulations" (Clinton and Gore 1995, 14–15).

The Clinton administration envisioned a system of quid pro quo in Project XL: If businesses achieved results better than would be achieved by full compliance with existing law, EPA would waive some existing requirements (Clinton and Gore 1995). Under the quid pro quo, the government would establish a "high bar" of environmental performance for excellent companies and would give them the flexibility to decide how they were going to "jump over it." The expected benefits included increased flexibility to adopt innovative solutions to environmental problems, increased (and more cost-effective) environmental protection, improved compliance, increased use of innovative technologies, expanded use of waste minimization and pollution prevention strategies, and a more cooperative relationship among regulators, the facility, and the community. For companies willing to meet the criteria for entry into this system, the government was willing to "throw out the rulebook," a notion that at the time was controversial and continued to be so throughout the administration of Project XL (Steinzor 1996). The Clinton–Gore strategy, however, spelled out no specifics beyond what has been cited. It left the details of implementation for EPA to work out.

The Aspen Institute's Concept of the Quid Pro Quo

The Aspen Institute had played a significant role in creating the concept of the quid pro quo (Ginsberg and Cumis 1996). The Aspen Institute had been formed to "foster candid exchange among people of diverse viewpoints." It held dialogues that were in areas of "vital concern to the nation." From 1993 to 1996, at its Colorado campus, it conducted four meetings a year on environmental issues. Representatives from corporations such as Intel, Merck, 3M, and Weyerhaeuser, government officials from EPA, federal, state, and local agencies, and representatives from environmental organizations such as the Natural Resources Defense Council attended. These organizations became central players in the efforts to implement Project XL.

The leaders of these organizations wanted to develop innovative answers to environmental challenges that had resisted solution in the past. Their charge was to increase the opportunities for win–win solutions. They were critical of the old ways of managing environmental issues. Conventional practices, though an improvement over growth at any price, relied primarily on end-of-pipe cleanup, which allowed economic entities to carry on business as usual. End-of-pipe treatment helped ameliorate damage to human health and environmental quality, but treatment came after the fact, when products had already been designed and the means of producing them had been put in place. The old ways did not reduce pollution at its source, a far better approach.

To achieve reduction at the source, the Aspen dialogue endorsed several principles. It supported the economists' view that the polluter should pay for the damage that pollution caused. This principle was a good start toward creating policies that were cleaner, cheaper, and smarter. Another important principle was that there was need for greater integration of control across different types of pollution—air, water, and land. Strict media-based regulation limited policymakers, discouraging them from considering innovative solutions and making imaginative trade-offs that could improve citizens' health and protect ecosystems at less cost to society. The success of innovative approaches also rested on greater coordination among the different levels of government—federal, state, and local—and increased collaboration among key stakeholders. Aspen participants recognized the need for businesses, levels of government, and environmentalists to work together to achieve more cost-effective solutions. If they continued to approach each other with suspicion and hostility, these solutions would be out of reach.

The recommendations of the participants in the Aspen dialogue are summarized in "The Alternative Path—A Cleaner, Cheaper Way to Protect and Enhance the Environment" (Aspen Institute 1996).[2] The participants agreed that, to change the status quo, EPA had to have the authority to experiment. The new approaches that they wanted to try would not replace the current system; rather, they would supplement it. The participants were hesitant to stray too far from the current system for fear that an extreme change might cause excessive harm to the environment or economic instability to firms. Under the new system, businesses and governments would have to operate with different principles and methods. The results of operating in this way were unknown, and the potential risks made it necessary to proceed with caution. The old and new systems therefore would run in parallel and the current system would remain in place and serve as a benchmark, so that regulators and businesses could determine if environmental performance in fact was superior. The Aspen participants agreed that the alternative path would be reserved for the best corporate actors, those with a record of responsible environmental performance and a zeal for environmental excellence. Only those organizations that could demonstrate that they were achieving a high level of environmental performance would be allowed into the new system.

Environmental policies based on closer business–government–stakeholder ties were needed, but beyond the idea that there should be a separate path, it was not clear to the Aspen participants how these policies should be put in place. Those involved in the dialogue decided that the answer should involve a quid pro quo—that in exchange for greater regulatory flexibility, businesses should demonstrate superior environmental performance (SEP). Under the quid pro quo, regulated businesses would take the initiative to show that they could achieve SEP through the design of special environmental management systems. These systems would be appropriate to circumstance and place and to an organization's capabilities for solving problems at their source without regard to media—air, water, or land. Firms would have to show that they were able to take advantage of pollution prevention and other innovative approaches, such as design for the environment, total product responsibility, total quality environmental management, and full cost accounting. The burden of proof was on the companies to demonstrate what they could do. The incentive for them to act was the operating flexibility that they would gain through the relaxing or the waiver of particular regulatory requirements.

According to Aspen participants, SEP meant achieving environmental outcomes better than simple compliance with applicable regulations. This definition was somewhat nebulous and presented certain problems to the participants, however. Less than allowable releases—a relatively straightforward benchmark—became more complicated when a facility already was operating below current requirements. Another standard suggested by the alternative path participants was achieving less than a facility's current, or actual, releases. Defining SEP as less than current or actual releases put a facility, which already was operating below current applicable requirements, in the difficult position of having to determine how much farther it could go. In some instances, the facility already had expended considerable resources and stretched its technical expertise to reach its current point. The participants in the Aspen dialogue recognized this problem, and, therefore, they held that the appropriate benchmark denoting SEP could be negotiated on a case-by-case basis. But they were not clear about how to conduct these negotiations.

The issue of how to conduct the negotiations posed a serious problem. Making complex SEP determinations when it was unclear how the negotiations should be conducted could delay the process. Companies were seeking greater flexibility and quicker decisions to respond to changing market conditions, not new constraints and more delays. The new approach, which the alternative path advocated, was supposed to enable facilities to flourish and to respond rapidly to market forces without undue regulatory impedance, while ensuring excellent environmental performance. Site-specific agreements had to be developed in an open, transparent way that involved a consensus-based stakeholder process. But without a clear demarcation of authority and substantial trust between the parties, they would not get far in their negotiations. There was no reason to depart from the status quo if a new approach was more time-consuming and burdensome than the existing system.

The Aspen dialogue recommendations of definitions of SEP were imprecise about which standard would be applied and when. Was it to be better than allowable releases, better than actual releases, or some other standard such as best performance practices or pollution improvements on the basis of units of production? The lack of a simple benchmark to measure SEP necessitated the use of decisionmaker judgement. But the Aspen participants also were vague about who would make the judgement calls, how they would be made, and on what basis. Ultimately, the participants in the alternative path suggested that discretion should be given to those who negotiated an agreement without being definitive as to what the appropriate benchmark for SEP would be. A key issue never confronted was who the negotiating parties would be and how they would work together.

The success of the quid pro quo depended not only on better environmental performance, but also on better economic performance. Businesses obviously were hoping to obtain competitive advantage through the flexibility they would achieve. But would the new approach actually yield the efficiency gains—the lower costs—that businesses were seeking? Business also faced a degree of risk and uncertainty not considered at Aspen. What happens if, after the application of significant resources to an innovative approach to environmental management, the desired SEP is not achieved? Does the company lose its operating flexibility and have to make new investments, such as costly, end-of-pipe controls? How much time should be allowed for SEP to be achieved? Perhaps the alternative path only got as far as it did because details such as these were sketchy and they allowed the participants to avoid disagreement.[3] These questions would plague the participants in XL pilots as they struggled with the difficult task of conducting experiments.

EPA Seeks XL Proposals

During this period, there was a commitment in many quarters to experiment and move beyond the limitations of the existing system (Bryner 1996).

Congress's Office of Technology Assessment, for instance, called for experiments with different approaches. In a 1996 report, it asked for state and regional experiments in using regulatory instruments with which the United States had little experience (Percival and Alevizatos 1997). It maintained that U.S. policies should become more results-oriented. Regulatory agencies should specify results rather than the means for achieving them so that sources would have greater flexibility to achieve the targets in ways that would be more cost-effective or otherwise beneficial.

In the same year, the President's Council on Sustainable Development (PCSD) also issued a report, "Sustainable America: A New Consensus—Building a New Framework for a New Century" (President's Council on Sustainable Development 1996). PCSD maintained that the nation had to go beyond the command-and-control structure that had proved effective in the past (Percival and Alevizatos 1997) and develop a wider array of approaches. The lessons of 25 years of command and control, according to the council, were that

- economic growth, environmental health, and social equity could not be addressed in isolation;
- the adversarial nature of the current system was counterproductive;
- collaboration and cooperation among past adversaries were needed;
- pollution prevention was better at encouraging technological innovation than technology-based standards;
- if compliance were assured, enhanced flexibility for achieving environmental goals was needed; and
- many of the most creative and lasting solutions came from partnerships and collaborations at the state level.

According to reports such as these, experiments were needed if regulatory agencies, businesses, and affected communities were to attain the knowledge necessary to move forward (Dorf and Sabel 1998). New approaches had to be tried to see if they could deal with the increasingly complex and difficult environmental problems that had to be confronted.

There also was support at various EPA offices—but it was by no means unanimous.[4] Many at the agency did not agree that existing regulations were a barrier to additional environmental improvement. In carrying out Project XL under the Clinton–Gore mandate, agency staff and leaders engaged in correspondence and face-to-face meetings with representatives from environmental advocacy groups.[5] The environmental groups wanted the agency to push hard for environmental benefits. Representatives of nongovernmental organizations (NGOs) questioned how EPA would implement XL. A group of environmental NGOs sent a letter to Gore in September 1995.[6] This letter—signed by representatives of 26 NGOs, including the World Wildlife

Fund, Friends of the Earth, and the Natural Resources Defense Council—warned EPA against making XL "an escape hatch" from "accountability to the public and to principles of environmental justice." Among the concerns the NGOs raised were the selection of XL participants, the legitimacy of local stakeholder groups, and the definition of SEP. EPA responded to these concerns by promising to seek very high levels of environmental benefit and environmental excellence in carrying out XL projects.[7]

However, EPA did not meet all of the NGOs' concerns. The NGOs expressed their apprehension that Project XL might undermine efforts already underway in the Common Sense Initiative (CSI), another reinvention program and one in which they had participated heavily. In its response—written shortly after an October 30, 1995 meeting with the NGOs—EPA tried to establish a clear demarcation between these two programs. In a letter, David Gardiner, the assistant administrator of EPA's Office of Planning, Policy, and Evaluation, said that the purpose of CSI was to recommend to EPA changes in environmental regulations, statutes, and programs.[8] The program relied on stakeholder dialogue and consensus within an industrial sector about what these changes should be. Project XL, in contrast, was meant to test new performance-based approaches to environmental management by obtaining empirical data from field projects and then evaluating the results. Gardiner emphasized the data-gathering feature and experimental nature of XL projects. He disagreed with the contention that CSI should select XL projects and approve XL final project agreements for facilities in the six industrial sectors where CSI was active.

The Federal Register Notice

EPA set out its goals for Project XL in a *Federal Register* Notice (U.S. EPA 1995b). They included:

- environmental performance superior to that which would be achieved through compliance with current and reasonably anticipated future regulation;
- cost savings and paperwork reductions for regulators and the affected firm;
- stakeholder support, innovative environmental management strategies, and multimedia pollution prevention;
- transferability, feasibility, clearly defined objectives, measures of success, and time frames;
- easily understandable information, including performance data, made available by participating firms to stakeholders;
- consideration of risk to worker health and safety; and
- cooperative relations between regulators, the facility, and the community.

Note EPA's wording in the initial clause above. According to the quid pro quo that the White House proposed, environmental performance had to be superior to that which would be achieved by "full compliance with current laws and regulations." EPA's definition of SEP was somewhat different. It was seeking environmental performance superior to "that which would be achieved through compliance with current and reasonably anticipated future regulation." By adding the phrase "reasonably anticipated future regulation" and eliminating the phrase "full compliance with current laws and regulations," EPA was setting the quid pro quo bar higher than the Clinton administration. Many companies focused on the administration's statement and missed the significance of this shift upward. These different definitions of SEP were to become a major point of contention in the implementation of Project XL.

EPA's XL managers omitted the words "by full compliance with" and instead chose the phrase "through compliance with current and future anticipated regulation" to make it clear that superior implied more than just doing better than was allowed by the regulations.[9] According to an EPA official, the agency was looking for performance that was not only better than that required by the regulations but better than what the regulations actually produce.[10] The burden on an applicant was to convince EPA that its proposed pilot could achieve performance superior to what otherwise would have been attained had its facility continued to operate under the existing requirements. This test was demanding. For instance, if a facility's emissions were on a downward trend under existing requirements, it might have to prove to EPA that it would do even better than it already was doing. Additional improvement of this nature might not be technically possible. Besides, it is always difficult to predict the future. There could be changes in the market and in the process technology or in science and the understanding of environmental risk, which would make it difficult to achieve XL targets.

Some EPA officials wondered why the agency should grant XL permits to companies that were *just* doing better than current regulations. They argued that just doing better than what the law requires was not superior performance, because in actuality almost all facilities operated at levels better than the law required.[11] The reality was that plant managers did not want to bump up against the limits of the law and risk being in violation, which would subject them to enforcement action. Consequently, facility managers usually left themselves a little slack by making sure that emissions were somewhat less than required. Thus, some EPA officials argued that facilities had to go beyond what the law otherwise would produce, not just beyond what the law currently required. In doing so, the benefits for the environment had to be greater than the status quo.[12]

Because of this deceptively simple phrase—superior environmental performance—the quid pro quo became a source of contention as the parties tried to forge actual agreements under Project XL. Negotiating the operational meaning of SEP on a case-by-case basis with a framework this vague, uncertain, and subject to interpretation would prove to be complicated and controversial.

Notes

[1]As confirmed in the Clinton and Gore document and within parts of the titles of other publications, the title of this book is not completely original.

[2]Although these ideas were not published until 1996, the people involved in formulating them started to meet in 1993, two years before XL's birth. They came from companies, including ones that submitted early XL proposals; from EPA; and from the national environmental groups that carefully monitored the proposals. These representatives were cognizant of the kind of thinking that went into Aspen's ideas, thinking that not only was found among Aspen participants but also was fairly widespread among people influential in formulating environmental policy in this period.

[3]Frameworks for negotiation often are vague, indefinite, and subject to interpretation.

[4]Based on the 1997–1998 interviews with EPA staff. See the Acknowledgements for a description of these interviews.

[5]Letter from Public Interest Representatives to Al Gore on the Common Sense Initiative Input on Project XL Proposals, San Francisco, Sept. 21, 1995; letter from Public Interest Representatives to D. Gardiner, assistant administrator of EPA's Office of Policy, Planning, and Evaluation (OPPE), Oct. 27, 1995; "NGO/XL Conference Call," memo from J. Kessler, regulatory reform director of OPPE to Public Interest Representatives, Oct. 31, 1995; letter from D. Gardiner to J. Hironaka, a Public Interest Representative from the Santa Clara Center for Occupational Safety and Health, (undated) 1995; letter from F. Hansen, deputy administrator of EPA, to J. Hironaka, Nov. 1, 1995; letter from D. Gardiner to J. Hironaka, Nov. 2, 1995; letter from D. Gardiner to J. Hironaka, Nov. 30, 1995; Public Interest Representatives 1995; letter from J. Kessler to A. Ronchak, Project XL coordinator at Minnesota Pollution Control Agency regarding stakeholders, Oct. 18, 1995.

[6]Letter from Public Interest Representatives to Al Gore on the Common Sense Initiative Input on Project XL Proposals, San Francisco, Sept. 21, 1995; letter from Public Interest Representatives to D. Gardiner, assistant administrator of EPA's OPPE, Oct. 27, 1995.

[7]In its discussions with the NGOs, EPA showed it was concerned about citizen lawsuits. There was also discussion about the legal mechanism to implement Project XL (Pedersen 1995).

[8]Letter from D. Gardiner, assistant administrator of EPA's OPPE, to J. Hironaka, a Public Interest Representative, on concerns regarding integrity of XL stakeholder processes, October 31, 1995. Also, letter from D. Gardiner to J. Hironaka on followup to one action item, Nov. 2, 1995; letter from D. Gardiner to J. Hironaka summarizing action items, Nov. 30, 1995.

[9]It is interesting to note that in the introductory summary of the *Federal Register* Notice of May 1995, EPA used a compromise phrase, "through full compliance with all applicable regulations," which was closer to the White House's original version than the one EPA actually adopted.

[10]1997–1998 interviews with EPA staff.

[11]Ibid.

[12]Ibid. The fallacy in this argument is that, regardless of the standard, facility managers will leave themselves a little slack. They would allow themselves some slack in an XL permit as well.

3

Conflicting Goals

A significant obstacle to achieving cooperative solutions is the different goals held by the parties involved. The Clinton administration's aim was to have 50 approved Project XL agreement within one year. The U.S. Environmental Protection Agency (EPA), applying the principles from its 1995 *Federal Register* Notice (U.S. EPA 1995b), selected eight candidates for the first round of Project XL on November 3, 1995. Among the proposals it chose were those from the Minnesota Pollution Control Agency (MPCA) and the 3M Company.[1] MPCA's proposal had been submitted on June 10, 1995, and 3M's proposals had been submitted on July 10, 1995.[2] In the negotiations that followed, 3M was the main actor on the business side. The main actors on the government side were MPCA; its stakeholder group, the Pilot Project Committee (PPC); and EPA. In this chapter, we introduce these actors. We show that they brought different perspectives, approaches, and goals to Project XL and that these differences led to difficulties in the negotiations that followed.

Business: The 3M Company

In 1995, when 3M was selected to be part of Project XL, the company had sales of more than $14 billion—about half outside the United States—and roughly 70,000 employees worldwide. 3M is a globally diversified manufacturer that sells thousands of products and services. Its headquarters are located in a suburb of Saint Paul, Minnesota. Among its more notable consumer products were Scotch® Cellophane Tape, Post-it® Notes and Flags, 3M™ Sandpaper, Scotch-Brite™ Scour Pads, and Scotchgard™ Protectors. In the United States, 3M operated in 33 states. In 1995, it was organized into more than 40 product divisions, subsidiaries, and departments, which competed in six markets: industrial; transportation, graphics, and safety; health care; consumer and office; electronics and telecommunications; and specialty materials. 3M's formal organizational structure had less meaning than its technology platforms (see Table 3-1), upon which it built many businesses for diverse markets.

3M believed that its advantage in competing with other companies came from the speed with which it introduced new products to the market (Gundling 2000). The regulatory system, where permitting can be a lengthy process, was a potential obstacle. Until 1993, 3M's goal was that 25% of its products should be no more than five years old. In 1993, it raised this goal to 36% being no more than four years old.

The 3M ethos and operating style help us understand the company. A plaque of a 3M memorandum that hangs in the office of many employees

Table 3-1. 3M 's Technology Platforms

Acoustics	Medical devices
Adhesives	Microbiology
Batteries	Microreplication
Ceramics	Molding
Coated abrasives	Nonwovens
Copper interconnects	Optical fibers and connectors
Dental and orthodontics	Optics and light management
Display materials	Particle and dispersion processing
Drug delivery	Pharmaceuticals
Electromechanical systems	Polymer melt processing
Electronics	Porous materials and membranes
Films	Precision coating
Filtration	Radiation processing
Fluorochemicals and fluoropolymers	Reclosable fasteners
Fuel cells	Skin health
Health information	Software
Imaging	Specialty chemicals and polymers
Infection prevention	Surface modification
Inks and pigments	Vibration damping

reads: "Those ... to whom we delegate authority and responsibility, if they are good ..., are going to have ideas of their own and are going to want to do their jobs in their own way ... mistakes will be made" (Stewart 1996). This quotation, from William McKnight, a former chief executive of the company, epitomized the company's dedication to innovation despite the risks. The company had been very successful in acting on the basis of this principle. Holding inflation constant, gain in real sales per employee from 1984 to 1995 was 101%, in comparison with an industry average of 39% (Stewart 1996). As of February 1996, 3M had been on *Fortune*'s most admired list 10 out of 11 years.

A Leader in Environmental, Health, and Safety Issues

3M also was one of the first companies to recognize the importance of environmentally responsible operations. In 1975, the company adopted an environmental policy, which said that it would:

- solve its own environmental problems;
- prevent pollution at the source whenever and wherever possible;
- develop products that have a minimum effect on the environment;
- conserve natural resources through reclamation and other appropriate methods;
- meet and maintain government regulations; and
- assist government agencies in environmental activities wherever possible.

Though 3M did not use the term *strategic environmental management*, it is a good description of what the company did. 3M's employees claimed that they refused to sacrifice the environment for the sake of corporate profits and that they tried to harmonize these ends. Their goal was to develop solutions that were good for both the environment and the corporate bottom line.

In 1975, 3M created the 3P (Pollution Prevention Pays) program, which encouraged employees to prevent pollution rather than treat it after it is generated. Under the leadership of its chief environmental officer, Joe Ling, 3M was an innovator in pollution prevention, developing the first successful industrial program and vigorously promoting the concept through the 1970s and early 1980s, before government and other companies became involved (Sinsheimer and Gottlieb 1995). The program's policies (3M Company n.d., 1990a, 1990b, 1991a, 1991b, 1991c, 1992) committed the company to source reduction by means of product reformulation, process modification, equipment redesign, recycling, and reuse.

3M was a moderating force in a business community that often opposed environmental initiatives without giving them careful thought or consideration. For instance, 3M belonged to the Business Environmental Leadership Council (BELC) along with companies such as Boeing, BP America, DuPont, and Toyota. BELC argued that living standards did not have to suffer in addressing global climate change. Emissions could be reduced, and the world economy still could thrive. This position was at odds with the thinking of many in the corporate community, who believed that the problem of global warming could be tackled only if living standards diminished. 3M had a long history of being an active and prominent member of associations like BELC.

Many 3M employees saw its environmental initiatives as an opportunity to display their ingenuity and problem-solving skills. The company's policies recognized the reality of threats to the global ecosystem and human survival, and they aimed to reduce or prevent pollution within the expanding global marketplace. They internalized and operationalized the principle of sustainability as a way to achieve environmental protection with economic growth. They endorsed and promoted eco-efficiency—making products with the least amount of materials and least amount of waste.

3M employees acknowledged that the advantages of a good environmental image were hard to quantify but felt that having this image helped the company reach customers (3M Company 1996a), and affected sales by differentiating its products and operations from those of its competitors. Environmental issues were also of particular concern to 3M because of its heavy use of volatile organic solvents in the production of many of its products.

In 1990, under the direction of its then–chief executive Al Jacobsen, the company decided to reinvigorate its 3P program. It established new goals to achieve by the year 2000: a 90% reduction in all chemical releases and a 50% reduction in waste. Its targets for the twenty-first century were to be as close to zero pollution as possible. The company proclaimed its wish to be ahead of regulation and to practice sustainable development.

To assure that its products were manufactured, distributed, used, and disposed of properly, 3M also established a program of full life-cycle environmental responsibility. The aim was to give careful scrutiny to the environmental impact of new products. Instead of reviewing only the manufacturing process, where most environmental regulation focused, 3M assessed a product's impact throughout its life cycle—from design to processing to distribution to use and, finally, to disposition. The goal for all new products was to minimize toxicity and energy use and to build in high reusability, ease of recycling, and safety. To an increasing extent, 3M was using the concept of full life-cycle environmental responsibility, and it had a growing assortment of products that it brought to market, or was considering bringing to market, that had been subject to this type of analysis. 3M claimed that its environmental policies had opened new markets and lowered costs.

3M also was implementing a new corporatewide environmental management system. By implementing this system, the company could assure consistent application of its policies. It was company policy that facilities outside the United States were subject to standards as rigorous as those applied in the United States, which included goals of a 90% reduction in emissions and a 50% reduction in waste. Each facility would have to install best available control technology, as defined by 3M environmental managers, on all sources that emitted more than 100 tons of volatile organic compounds a year. Each 3M facility was regularly assessed and reviewed to assure compliance with 3M policies and to ensure that it was demonstrating continuous improvement and movement toward sustainable development. 3M's policy was to comply with either applicable government regulations or company objectives, whichever was stricter.

The Saint Paul Permit

3M had a long history of working cooperatively with MPCA. In 1993, 3M and MPCA's Air Quality Division completed negotiations for a flexible permit at the company's adhesive tape plant in Saint Paul.[3] This permit set up a new, plantwide applicability limit, a so-called bubble over the Saint Paul plant that avoided source-by-source limits and replaced it with a single, overall plant requirement. The plantwide applicability limit gave 3M increased flexibility to operate without triggering permit modification requirements, thereby avoiding a lengthy administrative process. According to Tom Zosel, manager of environmental initiatives at 3M, the Saint Paul permit was among the first in the United States to set plantwide limits and provide some operating flexibility under these caps.[4]

The flexible permit for the adhesive tape plant in Saint Paul was based on emissions caps, installation and operation of voluntary pollution control systems, and more flexible construction and installation requirements. The permit granted operating flexibility to limited modifications in existing equipment only, but changes that could trigger the application of New Source Performance Standards were not allowed without the usual permit reviews and procedure. For example, the permit gave no regulatory flexibility for the introduction of new equipment, such as a new coater that applies adhesives to tapes, which would have to undergo the usual permitting process.

The Saint Paul permit benefited all parties. Volatile organic compounds (VOCs) went down more than 50%, from more than 10,000 tons a year to less than 4,300. MPCA saved about 730 hours of work. 3M made 21 facility changes without having to reauthorize the permit and saved approximately 1,500 hours of work. Though the Saint Paul permit served the parties well, it did have limitations. It was not multimedia in character. The sole focus was on air pollutant emissions. No new technologies were tested, and no innovation in pollution prevention took place. To achieve reductions in emissions in Saint Paul, 3M relied mainly on thermal oxidization, a pollution control method that removed emissions at the end of the pipe. The flexibility granted the facility was limited. The rapid introduction of a new product would be impeded by the need to submit to the usual permitting process.

Even so, the Saint Paul permit was a move toward a performance-based approach and away from source-by-source regulation. By setting a cap on total VOC emissions, subject to the constraint that there would be no violations of ambient air standards, plant engineers were free to improve overall environmental performance by upgrading existing production systems in the most cost-effective ways. As long as new production lines were not introduced and regulatory thresholds were not crossed, 3M could do this without experiencing the usual costly and time-consuming regulatory procedures.

3M applied for this type of permit because it anticipated problems and expenses in maintaining and upgrading its processes and equipment at the Saint Paul facility. There were 10 process changes needed that normally might require prevention of significant deterioration reviews under the Clean Air Act. To avoid or reduce the possible delays from these reviews, 3M agreed to cap the plant's emissions at the existing level, which at the time was well below Clean Air Act and MPCA requirements, and to install continuous emissions monitoring instrumentation on its VOC emissions sources. In return, the company had the right to modify processes and existing equipment (but not new equipment) without permit revisions and without prevention of significant deterioration reviews (Faegre & Benson 1994; Environmental Resources Management Group 1993a, 1993b; Air and Waste Management Association 1996a, 1996b). This innovative permit for 3M's Saint Paul facility was an important precursor of the more ambitious XL proposal the company made for its Hutchinson, Minnesota, plant under Project XL.

Project XL

3M came to Project XL with the intention of working with government to improve environmental performance while simultaneously achieving operating flexibility to bring new products to market more rapidly. Though personnel at the Hutchinson facility were involved, 3M's corporate environmental staff in Saint Paul spearheaded the company's efforts. Five staff members in the Environmental Technology and Safety Services Division were responsible for new initiatives and had the leadership role. At the head of this group was Tom Zosel, who vigorously, forcefully, and, at times, even abrasively championed 3M's XL proposal. Zosel was a veteran environmental manager and a well-known national leader in the effort to reform environmental regulation. He had been a participant in the Aspen Dialogue and had independently proposed his own version of an alternative-path, beyond-compliance concept. The other members of 3M's team came from the company's public affairs, legal, and technical staffs.

State Government: The Minnesota Pollution Control Agency

MPCA was established in 1967 to protect the state's air, water, and land from pollution. Its primary responsibility was to implement relevant federal laws and regulations, subject to EPA oversight.[5] MPCA had a staff of about 800 people. The governor appointed a Citizen's Board, consisting of eight members, to set agency policy and direction. Under authority delegated to MPCA by this board, the MPCA commissioner directed routine operations.

In 1991, Minnesota's Republican governor, Arne Carlson, appointed Charles Williams to be MPCA commissioner. Williams had been head of the Western Lake Superior Sanitary District in Northern Minnesota and a manager at Reserve Mining, a major producer of iron ore. In these positions he had come to know MPCA first-hand as a permittee. Williams' aim was to make the agency more efficient and effective. He instructed his staff to reduce adversarial relations, change the way regulation was carried out, streamline the permitting process, shift the environmental management paradigm, and make MPCA more of a partner with the regulated community.

Project XL

MPCA staff were pleased that EPA had named MPCA an XL participant because they viewed it as an expression of EPA's confidence in the state agency.[6] They interpreted it to mean they could experiment without elaborate federal supervision. Williams selected Lisa Thorvig to manage MPCA's XL pilots. Thorvig had been director of the Air Quality Division since 1991 and had spent her entire career at the state agency working on air quality and Superfund issues. To manage the XL process, she established an integration team, which included representatives from the major divisions in the agency. There were three people from air quality (including Thorvig), two from water quality, one from hazardous waste, one from solid waste, and a pollution prevention specialist from the office of administrative services. The team also had representatives from the Minnesota Office of Environmental Assistance and the Attorney General's Office.

During this time, much to Thorvig's dismay, not everyone in the agency supported Project XL.[7] Some resisted because they were concerned that XL would mean additional work. Others did not want to depart from a familiar regulatory structure, fearing it might lead to compromises on environmental protection. MPCA top managers tried to overcome this resistance by repeatedly explaining XL's philosophy and rationale to the staff.

MPCA's XL proposal (MPCA 1995b) called on the PPC of the Collaboration for a Better Environment and Economy (CBEE) to be MPCA's stakeholder group.[8] PPC was supposed to advise and oversee pilot projects in Minnesota, understand and evaluate their success in encouraging pollution prevention, and explore the ability to transfer the results of successful approaches (Marcus, Geffen, Sexton, and Smith 1995, 1996a, 1996b, 1997). PPC would work toward making the Minnesota experiment, if successful, a permanent option for as many companies and industries as possible, both in Minnesota and nationally. 3M had its own stakeholder group composed of citizens from the city of Hutchinson. The use of two stakeholder groups was a unique feature of the 3M project.

CBEE, of which PPC was a part, had been formed during the early 1990s with the goal of carrying out policy dialogues to explore the roots of environmental problems and develop novel solutions (Marcus, Geffen, Frisch, et al. 1997; Marcus, Geffen, Erickson, et al. 1998).[9] It sought to explore innovative ways to facilitate collaborative solutions to complex environmental problems by forming focus groups or roundtables. It hoped to:

- increase mutual understanding of the position and motivations of the different stakeholder groups,
- agree on a working definition of a problem and arrive at common objectives,
- foster feelings of goodwill and trust,
- agree on basic scientific and engineering facts and uncertainties,
- seek new ideas and approaches, and
- draft recommendations for changes in regulation and how business incorporates environmental considerations into decisionmaking.

CBEE's multistakeholder discussion, called the Pollution Prevention (P2) Dialogue, brought together participants from business, government, environmental organizations, and academia and examined the regulatory and business barriers to pollution prevention and the means available to overcome them. Lisa Thorvig had been an active participant in the P2 Dialogue.

The 10 members of PPC included three environmental managers from leading companies in Minnesota, two environmental engineering consultants, an environmental lawyer from a large Twin Cities law firm, an attorney with the Minnesota attorney general's staff responsible for environmental issues, two members of environmental advocacy organizations, and one of the authors (Ken Sexton) from the University of Minnesota School of Public Health. The other authors of this book served as PPC facilitators and participants in the discussions.

PPC members also were among the signers of the document "Recommendations of the Pollution Prevention Dialogue," which the participants in the P2 Dialogue had deliberated over and discussed for more than 15 months. This document committed them to encouraging source reduction and other innovative environmental management methods. Besides pollution prevention, PPC was interested in seeing the broader application and use of such other innovative solutions to environmental problems as design for the environment, life-cycle assessment, environmental accounting, total cost assessment, ecoefficiency, and strategic environmental management.

Regulatory Flexibility in Exchange for Beyond-Compliance Permitting

The ideas for 3M's XL project had been aired at one of the P2 Dialogue meetings. Representatives from 3M and MPCA presented a plan for having companies go 25% beyond compliance in exchange for greater permitting flexibility.[10] Similar to the beyond-compliance idea supported by 3M at the Aspen meetings (Aspen Institute 1996), the company's alternative to the current system was meant to be used by so-called good actors. Permit holders satisfied with existing regulation or having a poor compliance record would continue to operate under the existing system. Permit holders who had demonstrated superior performance, however, would have the option to choose an alternative that would grant them increased flexibility.

3M staff had drafted a Beyond Compliance Emissions Reductions bill that they hoped would be introduced in Congress (3M Company 1995a, 1995b). This bill stipulated that a facility could ask for a single, multimedia permit that had performance-based standards, which would replace the many individual technology-based permits a facility was required to have under current arrangements. The bill sought to reduce transaction time and costs associated with the implementation of environmental laws such as clean air and water acts and the Resource Conservation and Recovery Act. As long as a facility met the requirements of a beyond-compliance permit, plant personnel could make major modifications or introduce new processes and products without obtaining new permits from a state regulatory agency.

An important feature of the proposed law was 3M's suggestion to determine a facility's legally binding performance limits by making a calculation based on the "predominant medium" (air, water, or land). The predominant medium was defined as the medium experiencing the greatest environmental harm from the plant's releases. For instance, if that medium was air, then the facility would establish the plantwide total annual amount of air pollutant emissions it was allowed by regulations when operating the facility during any of the previous three years. It then would bind itself in a permit to a reduction in pollutant emissions of at least 25% below that amount. Releases into other media would be capped at the facility's actual total emissions level, assuming it was operating in compliance with all regulations. The plant could choose two main media and commit itself to a reduction for both of at least 20% below its allowed emissions, calculated in the manner described.

Another important feature of the proposed legislation was that the required pollutant emissions reductions were based on past performance. For example, a facility that had previously reduced its air pollutant emissions to more than 25% below what was allowed would be immediately eligible for a beyond-compliance permit and already would be in compliance. Thus, facilities that had made substantial progress in the past had an incentive to participate in this system. As 3M conceived it, the Beyond Compliance Emissions Reduction bill was meant to reward and encourage the behavior of companies that had shown in the past that they were environmental leaders.[11]

An additional feature under 3M's plan was that state governments, not the federal government, would have the authority to issue beyond-compliance permits. State governments would work closely with citizen groups and localities. The governor of a state that issued beyond-compliance permits would establish a local advisory council consisting of affected interests in the community where the facility was located and would seek advice and comments from this body during a review period. Thus, 3M's plan delegated substantial authority to state governments and local communities, a provision that in the context of prior pollution laws was very controversial.[12]

The Pilot Project Committee

In response to the MPCA–3M plan, the P2 Dialogue established a committee (the PPC) to explore the benefits of a beyond-compliance project.[13] The PPC's goals were to:

- maximize the incentives for pollution prevention in the permitting process,
- help make the pilot as broadly applicable to other industries as possible,
- use the diverse expertise and experience of P2 Dialogue participants as a resource for pursuing a pilot,
- help mobilize support for the pilot,
- provide a litmus test for potential opposition to the pilot and provide remedies to valid objections,
- track outcomes at participating organizations,
- estimate the extent to which pollution prevention solutions were stimulated, and
- observe the long-term effects.

Concern about the role that pollution prevention would play in a pilot, however, led to a dispute about the meaning of pollution prevention. The dialogue's issue paper defined *pollution prevention* as reducing potential pollution at input stages rather than at the output stage. When new state and federal legislation was passed at the end of the 1980s and the beginning of the 1990s, the environmental community argued for a definition of pollution prevention as source reduction only, but elements in the business community believed that prevention should include additional options, including recycling and reuse. This controversy resurfaced during the dialogue, and despite the efforts the facilitators made to mediate it, 3M's representative became irritated and the company dropped out of the dialogue.

The final recommendations of the P2 Dialogue appeared in a report (CBEE 1996). They resembled the recommendations of other groups like Aspen (Aspen Institute 1996) in a number of ways: dissatisfaction with the existing regulatory system, calls for greater flexibility, arguments in favor of experiments with alternative approaches, and support for outcome-based regulation. The report also asked for new and meaningful ways to engage stakeholders (Marcus, Geffen, Erickson, et al. 1998).

The P2 Dialogue emphasized management methods that reduced environmental impact and costs and encouraged innovation. It viewed the current regulatory system as playing an important role but also as an obstacle that had to be overcome before further progress could be made. The P2 Dialogue urged government, industry, public interest groups, and other stakeholders to work together to increase incentives and lower barriers to greater source reduction and other innovative environmental management methods. It called on government to spur innovative solutions to pollution problems at lower costs by establishing outcome-based requirements that the regulated community could meet in ways best suited to its circumstances. The dialogue also called on the regulated community to promote innovative management techniques such as design for the environment, life-cycle assessment, and environmental accounting systems that would better measure the environmental costs of producing products and the environmental costs to consumers of using them. The P2 Dialogue asked both regulators and the regulated community to engage in discussions with interested stakeholders and the local community early in any process involving significant changes in permitting.

Enabling Legislation

From June 1995 to January 1996, MPCA worked with its stakeholder group (the PPC) and 3M to frame enabling legislation. This legislation provided MPCA with the legal authority it needed from the state government to conduct regulatory experiments and to suspend, if necessary, existing Minnesota environmental statutes in order to test new ideas (PPC, notes from monthly meetings with MPCA, 1996–1997).

To pass the Environmental Regulatory Innovations Act (Minnesota State Legislature 1996), MPCA had to overcome the skepticism of environmental organizations and concerned legislators. The language of the bill was modified in response to their criticisms. The commissioner wrote a special letter of commitment to the environmentalists, which promised that the pilot programs would experiment with multistakeholder, consensus-based decision-making and guaranteed that there would be no increase in biopersistent compounds. The legislation, passed in March 1996, gave an added boost to the concept of pollution prevention and provided MPCA with extensive authority to carry out regulatory experiments. As of this writing, though bills have been proposed, the federal government has not seriously considered passing similar legislation.

Federal Government: The Environmental Protection Agency

EPA was established as an independent agency on December 2, 1970. It was created through Reorganization Plan 3 of 1970, which consolidated federal environmental regulatory programs into a single agency by bringing together 15 components from five executive departments and independent agencies. EPA's mission was to administer the various statutes over which it had authority.[14] To do so, it was organized into 13 headquarters offices (12 of which were in Washington, D.C., and one at Research Triangle Park, North Carolina). These offices were:

- Office of the Administrator;
- four regulatory program offices (Air and Radiation; Prevention, Pesticides, and Toxic Substances; Solid Wastes and Emergency Response; and Water);
- seven offices with responsibilities that cut across regulatory programs (Policy, Planning, and Evaluation; Enforcement and Compliance Assurance; Finance; General Counsel; Inspector General; International Activities; Environmental Information); and
- Office of Research and Development.

EPA also had 10 regional offices with responsibility for carrying out existing environmental laws within the boundaries of their multistate regions. Each region has a somewhat different history and culture, which affects how they treated Project XL proposals. The regional offices had parallel structures to headquarters with functions such as enforcement and air and water quality. The regions, however, did not have a research and development capability, nor did they have responsibility for policy or for creating regulations (see Marcus 1991).

Groups within EPA that played an important role in the 3M–Hutchinson XL project included the agency's regulatory reform staff in Washington, D.C., the Office of Air Quality Planning and Standards (OAQPS) in Research Triangle Park,[15] and the Region 5 Office in Chicago.[16] The regulatory reform staff initially was located in the EPA Office of Policy, Planning, and Evaluation (OPPE) but later it moved to the newly created Office of Reinvention, which then became the Office of Policy, Economics, and Innovation (OPEI).[17] OPPE worked very closely with the EPA Office of General Counsel, which was concerned about legal authority for implementing the pilots as well as defending them against challenges in court. Originally, when the administration of Project XL was in OPPE, David Gardiner, the assistant administrator of OPPE, assigned Project XL to Jon Kessler, director of the Emerging Sectors and Strategy Division. Kessler pulled in people representing the various programs and other parts of EPA. Very often, they were the same people who had been working with the White House on the reinvention effort.

When considering participation in Project XL, the first contact an organization would have was likely to be with EPA headquarters (that is, OPPE). If the pilot was accepted into XL, management of the project then would be delegated to the appropriate region. Thus, EPA regions had little input into EPA's initial policies, becoming involved only when pilot project proponents came forward to talk to EPA or submitted their ideas about proposals to the agency. The regions, however, were to manage the projects; the applicable regional administrator would have explicit responsibility for project development and was responsible for bringing policy issues of national consequence to the attention of senior EPA officials. When XL started, it was assumed that such issues would arise and that OPPE would have to develop a working consensus among the different national offices to resolve these issues. Little was done initially, however, to develop a practical management structure that could achieve this consensus in a timely fashion.

The main participants in the 3M–Hutchinson project from Region 5 were from the Office of the Regional Counsel, the Air Quality Division, and a small coordinating group from the region's Office of Strategic and Environmental Analysis. Because the region had responsibility for reviewing state implementation plans, it took on an advisory role and examined the legal issues involved in proposed XL pilots. The region expected substantial discussion of legal issues before pilot projects were approved, and a staff person from the Office of Regional Counsel was appointed to direct the region's XL team.

Participants' Goals with Reference to Project XL

The goals for Project XL of 3M, MPCA and its stakeholder PPC, and EPA were not fixed but rather evolved, becoming more or less clear depending on the circumstances and the particular individuals who articulated them. Each of the groups involved was a multidimensional organization with various subgroups and factions that did not always agree. The goals of each group tended to change from situation to situation, sometimes substantially. The discussion below briefly summarizes the different and changing goals of the 3M–Hutchinson XL participants. Our understanding of their evolving goals is based on our participation in the project, extensive interviews with the main actors, and a review of the available documents. It is important to remember that there were subtleties as well as complex twists and turns in the elaboration and maturation of these goals that cannot be captured by this short summary. Most complex of all was EPA, where it is hard to say whether the agency had a cohesive and consistent set of goals for Project XL.

3M Company

3M had a number of reasons for proposing the 3M–Hutchinson XL pilot. Its main reason for seeking an alternative permit at Hutchinson was to gain competitive advantage from a flexible, performance-based, and facilitywide approach. 3M also wanted to test its beyond-compliance approach to regulation. In addition, the company saw Project XL as an opportunity to showcase its progressive initiatives, which benefited both the environment and the company's bottom line. 3M's primary purpose of achieving greater operating flexibility stemmed from its business needs for rapid process changes to meet shifts in the market. The company wanted to achieve fast cycle times in new product introductions and, to the extent possible, realize profits from being the first or nearly the first to the market. It did not want to be held back by regulatory barriers that could delay the process for months or years or, perhaps, derail it completely.

For those in environmental management positions at headquarters, XL would enhance their ability to put 3M's principles into practice: lowering waste to increase yields, minimizing environmental harm through life-cycle planning, and promoting sustainable development. 3M staff hoped that because of Project XL, fewer engineers would be involved in basic compliance activities and more would become involved in product responsibility and other new initiatives. The engineers' enthusiasm was for innovation and sustainability, not compliance and paperwork. The company's staff complained that they had little time left over after satisfying reporting requirements to work on a backlog of innovative environmental projects. The company also was active in researching new, more environmentally friendly production processes, and these efforts were slowed by the need to obtain permits to operate small production runs that tested new ideas and approaches.

The company's managers and engineers argued that environmental principles and values were deeply embedded in the company, as evidenced by the voluntary reductions it made before Project XL. The employees cared about the environment, and they conceived of the XL project as granting them greater freedom to do "the right thing." They were adamant that they would continue to do what was right regardless of whether the company obtained a new permit under Project XL. Nevertheless, the staff had hopes that XL would serve as an important bridge to new and better relationships with regulators.

The Minnesota Pollution Control Agency

MPCA's goal was to conduct innovative experiments in regulatory reform using the platform provided by Project XL (Buelow 1996, 1997; Dews 1997). Its aim was to make the state's regulatory programs more efficient, responsive, user-friendly, and less costly. High on MPCA's list of objectives were cost savings and paperwork reductions for the agency as well as for regulated entities.[18] MPCA did not want to continue to use the bulk of its resources to maintain routine permits with companies that historically had acted in good faith by doing more than the law required. It wanted to ease the regulatory burden on so-called good actors like 3M, thereby freeing up resources with which to tackle previously underemphasized complex environmental problems, such as agricultural pollution. Declining state budget allocations and the burden of dealing with nonpoint-source pollution caused MPCA to be concerned that it lacked the resources to cope effectively with twenty-first-century environmental problems.

In addition, MPCA's management hoped that an innovative, streamlined capability in environmental management would improve the competitiveness of Minnesota businesses—an important concern for state governments. MPCA's commissioner and top managers were highly committed to the idea of replacing the current system with a more cooperative and cost-effective one. This support of environmental regulatory reform was not shared, however, by all MPCA staff, some of whom preferred the status quo.

The Pilot Project Committee

The main goal of PPC, MPCA's stakeholder group for XL, was to help develop a new approach to environmental management. PPC members supported the diffusion of various innovative environmental management strategies and techniques, such as pollution prevention, in the belief that adoption would benefit both the environment and the economy. The idea was to use Minnesota's XL projects not as a series of separate pilots, but as a broad demonstration designed to show the feasibility of implementing new, enlightened approaches to pollution control. The goal was to ultimately persuade many environmentally responsible companies to take an alternative pathway to conventional command-and-control regulation.

The U.S. Environmental Protection Agency

Since 1970, Congress had continually reduced EPA's regulatory discretion and had tightened relevant statutes to ensure that implementation would occur as Congress intended (Canon 1999; Bryner 1996; Davis 1977). The many congressional committees that had some jurisdiction over EPA were not reluctant to reinforce this message. Under the administrations of Ronald Reagan and George H.W. Bush, more emphasis was placed on evaluating the efficiency and cost-effectiveness of EPA's regulations and the burdens that regulations imposed on industry. During the 1980s, environmental litigation soared, as environmental groups sought to compel the agency to comply with existing laws. Consequently, EPA was cautious about exercising discretion and potentially deviating from the letter of the law. Many EPA staff members believed that strict adherence to statute was a defense against those who wanted to enfeeble the nation's environmental programs.[19] Allegiance to the rule of law was a significant element in the agency's culture that made it difficult for EPA to bargain with industry and environmental groups, even when such bargaining might bring about significant environmental gains at reduced costs.

Thus, EPA staff members were not unanimous about the need to move away from command and control and toward new approaches like Project XL.[20] Some staff, both in the regional offices and at headquarters, resisted changes to the current system. They believed that conventional approaches had worked for more than 20 years and should not be tampered with. Others wanted to move toward a revamped system because they saw the deficiencies with command and control and understood that it would not be effective in dealing with twenty-first-century environmental challenges such as global climate change and biodiversity loss. Reinvention was a controversial and contentious issue in the agency.

The internal debate crossed all the many reinvention programs the Clinton administration had proposed (Clinton and Gore 1995). Controversy arose within the agency whenever the details of any specific reinvention project had to be worked out. The controversy almost always had to do with the argument that EPA should adhere first and foremost to the existing body of environmental law and regulation and that it should not stray from it simply because of an administration's pet projects. However, the controversy also extended to basic values about what came first, the environment or the economy, and how these issues could be reconciled with EPA's congressional mandate to serve as a protector of environmental quality and human health.

At the top level in EPA, the Office of the Administrator and Deputy Administrator, staff understood reinvention to be a series of related initiatives whose objective was to gradually and carefully transform the agency and the regulatory system.[21] Fred Hansen, EPA's deputy administrator, expressed his concern that there was a widespread misconception that EPA's goal was more radical: to change the system in a fundamental way. Though the existing regulatory framework was prescriptive by nature, he believed that it was necessary. In the 1970s, when first established, Congress had mandated it, and the public had come to expect it. Hansen disagreed with critics who argued that some of the regulations the system imposed were silly. He thought that what critics regarded as unnecessary hurdles in fact provided important environmental benefits. According to Hansen, outmoded and useless regulations were not that common and were easily handled through the procedures that EPA already had in place.

Hansen recognized that many initiatives for change came from rapidly growing sectors of the economy (for instance, microprocessors), which were finding it hard to adapt to ponderous, slow-moving regulations—but not exclusively. He noted that he also saw these initiatives coming from older and slower-growing industries (such as metal plating) that traditionally had been hostile to regulation but now were more forthcoming with positive and constructive initiatives. Hansen believed that the climate for broadly based, systemwide change had been set back by the actions of the 1994 Congress.[22] Its attacks on the status quo and its proposals to fundamentally alter what EPA did made the agency's executives and leaders more suspicious and combative. A new course could be charted, but Congress made this task difficult.

EPA's leaders would not sacrifice citizens' rights to environmental protection for the kinds of reforms that congressional leaders—under the Contract with America—were discussing. Reinvention could be justified only when it assisted the agency in meeting the goal of environmental improvement. If reinvention established a tool or a new and better framework for the next generation of environmental protection, the agency favored it. If reinvention involved stakeholders and communities in a new way, EPA was behind it. If reinvention made regulation more transparent and open, the agency approved. EPA's top managers favored reinvention that permitted the agency to make additional environmental progress in a more commonsense, cost-effective manner. To EPA's leaders, reinvention did not mean wholesale restructuring and abandonment of past efforts. It meant acquiring new skills in the anticipation that they would strengthen the current system.

Proceeding with Caution

Thus, the signal EPA staff members were getting about conducting reinvention pilot projects was to be cautious and to go slowly. The choices that would have to be made were difficult and contentious ones. Hansen indicated, for instance, that there were likely to be substantive and legitimate differences about whether advanced approval for an overall pollution limit could be given in a particular setting. Honest disputes could arise about the amount of pollution involved, the capping of these amounts, the trading-off of different sources of pollution against each other, and the combining for regulatory purposes of the emissions from different sources.

The conflicts within the agency about reinvention were deep and fundamental, and EPA's consensus-based decisionmaking and management structure meant that nearly everyone both inside and outside the agency had a right to have a say. To resolve difficulties in EPA's policies, EPA's top leaders, including Hansen, were proponents of stakeholder involvement. Stakeholder participation was needed to help overcome obstacles faced in the specific pilots, but EPA's leaders had not determined precisely how stakeholders should be involved, what role they should play, or how they should play this role.

EPA's leaders understood that they had to satisfy competing interests (Marcus 1980). They were held accountable by Congress to enforce the many federal environmental protection laws. They also were held accountable by the White House to pursue the administration's reinvention agenda. Staff at the agency had a wide spectrum of attitudes, values, and interests. Significant and influential people, particularly policy analysts and economists, often were proponents of regulatory reform. They were among the most knowledgeable about regulatory reform issues and among the most committed to the goal of creating new approaches to environmental protection. The agency's main preoccupation, however, was with its statutory obligations, regardless of the associated costs, though EPA could not completely ignore the legitimate economic concerns of businesses. On the basis of a series of Executive Orders, the Office of Management and Budget in the White House had imposed this obligation on EPA—the agency had to consider the costs and benefits of important regulations (Marcus 1980). Consequently, it had to take into account jobs, economic growth, and the impact of its actions on industry survival. These pragmatic considerations played a role because of the Office of Management and Budget and the statutory constraints under which EPA had to operate. For example, technology-based approaches such as "best available control technology" included implicitly the understanding that "best available" meant at a reasonable cost.

EPA also had to balance its efforts at reinvention with its need to obtain political support. On the one hand, many in the business community complained about what they perceived to be excessive and overly costly regulations—regulations that, in their view, were not always justified by the available scientific evidence and by cost–benefit analysis. Increasingly, companies, even those like 3M that were committed to ambitious environmental performance goals, expressed their concern that existing regulatory controls impeded their ability to respond rapidly to changing market conditions. On the other hand, the environmental advocacy community had many people who were wary of these complaints. They feared that EPA's reinvention efforts would weaken environmental laws and increase the risk of harm to public health and the environment. The business community's views were backed by powerful lobbyists, who at times could count on the support of labor unions interested in preserving jobs and employment. Buoyed by a large measure of public support, environmental advocacy organizations also had influence, especially through the courts, which they had been able to effectively use when they believed environmental laws were not being properly enforced.

Because EPA operated in this setting of conflicting interests, its goals with respect to Project XL were hard to characterize precisely. Clearly, to the extent possible, the agency wanted to ensure that individual pilots achieved superior environmental performance. By "superior," EPA meant environmental performance or results that went beyond what would be achieved under current or reasonably anticipated future regulations, a judgement the agency would make by looking at a facility's past experience to determine what its environmental performance would be if it continued to operate under these requirements. To be successful in obtaining regulatory flexibility, a company that applied for an XL permit had to convince the agency that it could meet or exceed this threshold. In adhering to this principle, EPA's aim was to assure the integrity of existing environmental laws, while trying to accommodate the need for reform. In its deliberations, the agency therefore tended to focus on a central question: Did a proposed pilot project produce better results for the environment than otherwise would be produced by the regulatory framework already in place?

Even assuming that EPA would be able to initiate a meaningful number of worthwhile pilot projects that met this standard, many in the agency's top management were reluctant to use successful pilots to foster additional change in the regulatory system. Hansen, for example, expressed the view that EPA was seeking to learn lessons that could be applied to improve the existing system, not change it.[23] This approach was the one that EPA had taken previously when it experimented with facilitywide pollution caps (bubbles) and market-based trading. In EPA's view, reinvention initiatives might provide new tools that could augment and improve the existing system, not replace it with something new or different that might compromise the environmental protection the agency was obligated to preserve.

EPA did not share the view of the Minnesota participants—3M, MPCA, and PPC—that the need for change was urgent and that Project XL could be the start of a movement toward an alternative approach to environmental management.[24] The Minnesota participants saw Project XL as the beginning of a new system of environmental protection, one which would better serve all parties in the twenty-first century, whereas EPA saw it as more limited and having less potential. This disparity affected the participants' interactions during the negotiations for the proposed XL pilot at Hutchinson and was an important obstacle to reaching an agreement.

Notes

[1] The other six proposals were from Intel and Merck (both discussed in Chapter 8), Lucent Technologies (a part of AT&T at the time), Anheuser-Busch, HADCO (now a part of Sanmina–SCI), and the South Coast Air Quality Management District in California. Additional XL proposals by Weyerhaeuser Company (see Chapter 8), IBM, and Union Carbide were accepted shortly thereafter.

[2] The 3M proposal covered facilities in Bedford Park, Illinois, and Camarillo, California, as well as its Hutchinson, Minnesota, facility. In time, 3M withdrew its projects for the Bedford Park and Hutchinson facilities; the Camarillo facility was later ceded to Imation and spun off by 3M (see Chapter 4). Of the other eight initial proposals, the South Coast Air Quality Management District and Union Carbide withdrew from the program early on. HADCO terminated its XL project several years later after its acquisition by Sanmina Corporation, and the Anheuser-Busch project was not implemented.

[3] Pilot Project Committee (PPC), notes from presentation of 3M's plans for XL permit at Hutchinson, Feb. 29, 1996. Tom Zosel, manager of environmental initiatives at 3M, was the main 3M presenter; see Minnesota Air Emission Permit, 23GS-93-OT-1, 3/4/93.

[4] Ibid.

[5] Minnesota's authority to implement federal environmental laws is delegated to it by EPA and the state's performance is reviewed by EPA Region 5. If EPA is seriously dissatisfied with MPCA's administration of federal laws and no remedy is obtained, EPA could revoke MPCA's authority and Region 5 would become the regulatory authority for the state.

[6] Based on 1996–1997 interviews with MPCA staff. See the Acknowledgements for a description of these interviews.

[7] Ibid.

[8] As we earlier indicated, the authors of this book were involved in the CBEE as creators, facilitators, and participants. We were operating under a grant we received from the Joyce Foundation.

[9] The main participants in CBEE's creation were the Minnesota Environmental Initiative, a nonpartisan educational organization that brought together groups in collaborative forums to facilitate solutions to environmental problems, and the Strategic Management Research Center at the Carlson School of Management, University of Minnesota, which did problem-driven, interdisciplinary research that was of interest to private and public managers. Also involved in CBEE was the Center for Global Change (now defunct) at the University of Maryland, which carried out research on environmental and energy issues.

[10] CBEE, 3M's "Beyond Compliance," and MPCA's Pilot Project for Flexible Permitting, notes from Feb. 24, 1995, P2 Dialogue; presentation by D. Wefring, environmental regulatory specialist at 3M, "A New Paradigm in Environmental Permitting" (overheads), 1995.

[11]The new system would keep pace with new laws or regulations. The emissions levels to which a facility with a beyond-compliance permit was bound were not fixed for all time. They could be adjusted upward or downward depending on changes in statutory and regulatory requirements.

[12]A final notable feature of the proposed legislation was the way it encouraged pollution prevention and extended responsibility for product life cycles. A company could earn credit for helping its customers and suppliers reduce waste. The boundary between the company and those with whom it regularly dealt was relaxed in an effort to encourage pollution prevention and extended responsibility for products.

[13]CBEE, 3M's "Beyond Compliance," and the MPCA's Pilot Project for Flexible Permitting, notes from Feb. 24, 1995, P2 Dialogue. The authors of this book facilitated this effort and participated in it.

[14]Since EPA's inception, the enactment of major new laws and amendments to older laws have added greatly to its responsibilities (Marcus 1980; Landy et al. 1990; Fiorino 1995; Rosenbaum 1998). The agency administers many major environmental protection laws including the Clean Air Act; the Clean Water Act; the Safe Drinking Water Act; the Comprehensive Environmental Response, Compensation, and Liability Act (known also as Superfund); the Resource Conservation and Recovery Act; the Federal Insecticide, Fungicide, and Rodenticide Act; the Toxic Substances Control Act; and others. Less than a third of EPA's budget was committed to its operating programs (more than two-thirds went to water infrastructure projects and to trust funds), and the overall budget grew little in the 1990s.

[15]OAQPS was part of the Office of Air and Radiation. Its primary mission was to preserve and improve air quality in the United States. To accomplish this, OAQPS compiles and reviews air pollution data, develops regulations to limit and reduce air pollution, assists states and local agencies with monitoring and controlling air pollution, makes information about air pollution available to the public, and reports to Congress on the status of air pollution and the progress made in reducing it.

[16]The EPA Region 5 Office has oversight authority for the pollution control programs in the states of Illinois, Indiana, Ohio, Michigan, Minnesota, and Wisconsin. The office's responsibilities include cooperating with federal, state, interstate, and local agencies, industry, academic institutions, and other private groups to ensure that regional needs are considered and federal environmental laws are implemented; developing, proposing, and implementing regional programs to conduct comprehensive and integrated environmental protection activities, to translate technical program direction and evaluation into effective operating programs, and to assure that such programs are executed efficiently; exercising approval authority for proposed state standards and implementation plans; and evaluating overall and specific regional programs.

[17]In February 1997, EPA grouped together its regulatory reform staff in the new Office of Reinvention within the Office of the Administrator in Washington, D.C., and responsibility for Project XL was transferred to this office. Later, it was transferred to OPEI. The mission of the Office of Reinvention was to promote innovation to achieve greater and more cost-effective environmental protection. The associate administrator for reinvention served as the principal advisor to the administrator on matters pertaining to reinventing EPA's programs. The office was involved in a range of activities, including serving as a primary gateway for stakeholders and customers to use in interacting with EPA on reinvention; defining the vision, strategy, ground rules,

and principles for reinvention by engaging stakeholders; ensuring new approaches were identified, designed, and piloted by fostering program-specific approaches in other EPA offices and managing agencywide approaches; integrating and coordinating new approaches across the agency into a coherent strategy for change; tracking reinvention progress and evaluating reinvention success; and ensuring that successful new approaches were incorporated into the way EPA did business.

[18]MPCA 1995a, 1995b; also 1996–1997 interviews with MPCA staff.

[19]Based on 1996–1997 interviews with EPA staff. See the Acknowledgements for a description of these interviews.

[20]Ibid.

[21]Interview with F. Hansen, deputy administrator of EPA, February 11, 1997.

[22]Ibid.

[23]Ibid.

[24]See D. Geffen's summary of positions taken delivered at Project XL Minnesota Lessons Learned for Moving Forward Meeting, Minneapolis, April 1, 1997.

4

Complicating Factors

We have discussed the problems in the Project XL framework, the quid pro quo, as an obstacle to achieving a cooperative agreement and have examined the differing perspectives and goals of the participants. Now, we turn to another issue, a series of complicating factors—technical, economic, and legal in nature, some of which arose unexpectedly and all of which affected the ability of the parties to reach agreement. 3M Company staff, for instance, had to address complex technical issues relating to environmental performance at the Hutchinson facility. The flexible permit, which the company was seeking, covered an uncertain future in which 3M envisioned introducing new products and processes that would raise production at Hutchinson, while the company would have to keep pollution in check.

The company, moreover, experienced a change in economic fortune, a downturn at a critical juncture in drafting the agreement, which led to a corporate restructuring that was unprecedented in 3M's history. In addition, the U.S. Environmental Protection Agency (EPA), the Minnesota Pollution Control Agency (MPCA), and 3M had to proceed without clear answers to whether the project could be carried out without violating existing environmental laws. A final issue was MPCA's relationship with EPA Region 5. Did MPCA have the authority to issue an XL permit on its own?[1]

Technical Environmental Protection Issues
at the Hutchinson Facility

3M's Hutchinson facility is located in the southeast section of Hutchinson, Minnesota, a city with a population slightly under 12,000 that is about 60 miles west of Minneapolis. The facility consisted of two units, a North and a South plant.[2] The North Plant made audio and video (AV) tapes, including videocassettes and high-quality master reels for movie studios. The South Plant manufactured a variety of adhesive tape products, such as transparent, electrical, and masking tapes, and other products that had adhesive parts.

3M considered Hutchinson to be its flagship facility. The original plant site, acquired by the company in 1945, had experienced considerable growth. The close to 2,000 employees at Hutchinson enjoyed good wages and benefits. They had a strong work ethic, were among 3M's most experienced workers, and were committed to the company. They had well-honed skills in such core technologies as coating, extruding, compounding, grinding, slitting, winding, packing, and molding, which contributed to the high quality of the products they made. 3M had refined its processes over a long time, and few companies in the world could match the features of the adhesive tapes it produced at Hutchinson. The operations at the plant were proprietary so that 3M could protect its competitive edge. The company maintained its industry-leading position by making continuous improvements and distinguishing its products and processes in ways that prevented competitors from taking market share. Nevertheless, 3M recognized that other companies, with equally strong brand names and positions in their markets, had stumbled and that it could not afford to be lax. In a highly competitive market, domestic and foreign competitors were prepared to step in if 3M slipped.

Hutchinson was located in a designated "attainment area" (i.e., an area that had achieved air quality goals as specified in the Clean Air Act for six major air pollutants, referred to as criteria pollutants), so the facility was allowed to increase its pollution emissions as long as applicable regulations were satisfied. Nonetheless, according to Toxic Release Inventory data, more chemicals were managed (emitted, stored, and transported) at 3M–Hutchinson than at any other facility in Minnesota (see Table 4-1).

Though Hutchinson was the state's largest emitter of volatile organic compounds (VOCs), it had an excellent compliance record. At the time the XL permit was under development, the facility had more than 20 environmental permits. Emissions of VOCs resulted from the use of organic solvents to coat solids on a substrate, a critical step in the tape-making process. The central feature of the North and South plants was the tape production lines, called coaters, which applied the adhesive or magnetic coatings to the uncoated plastic tape. 3M dissolved magnetic or adhesive materials in organic solvents, such as methyl ethyl ketone (MEK) and heptane, coated the

Table 4-1. Chemicals Managed at Various Facilities in Minnesota, 1996

Facility	Quantity (pounds)
3M–Hutchinson	32,420,324
3M–Cottage Grove	14,301,073
Boise Cascade	11,094,610
Potlatch	5,452,925
North Star Steel	4,298,852
3M–Saint Paul	4,072,774
Ashland Petroleum	2,858,582
Koch Refinery	2,616,764
Sheldahl	2,504,580
Mixon	1,960,112

Source: Minnesota Office of Environmental Assistance, *1998 Pollution Prevention Evaluation Report.*

substrate, and evaporated the solvent in ovens.[3] Though other areas at 3M–Hutchinson yielded emissions (the mixing areas, parts washing, and solvent storage), the main sources were the five adhesive coaters at the South Plant and six AV coaters at the North Plant (MPCA 1996c).[4] 3M had been aggressively working to introduce water-based solvents into the tape production processes, but all of the magnetic tape coaters and many of the adhesive tape coaters still required the use of VOCs as solvents (MPCA 1996c). Many of these VOCs were also designated as hazardous air pollutants (HAPs).

Grandfathering

The federal Clean Air Act "grandfathered in" (i.e., allowed to operate without regulation) production facilities that had been in existence before the act was passed (U.S. EPA 1996f).[5] Only two of the Hutchinson facility's magnetic tape coaters and only one of the adhesive tape coaters came under the act's requirement to meet New Source Performance Standards (Faegre & Benson 1994; Environmental Resources Management Group 1993a, 1993b; Air and Waste Management Association 1996a). The remaining lines were not regulated. Other sources at both plants, primarily areas where coating materials were mixed (called compounding areas), also were unregulated, which meant that these production units could operate without pollution controls. However, new maximum achievable control technology (MACT) standards, prescribed by the Clean Air Act Amendments of 1990 to control emissions of HAPs, were due to take effect for the magnetic tape plant some time during the next few years, when the proposed XL permit would have been in force (U.S. EPA 1994).

The MACT standard established in the 1990 Amendments was defined for each manufacturing category as achieving control efficiency not less than that achieved by the best-performing 12% of existing controlled sources. Consequently, the new MACT standards would require all major sources at the magnetic tape plant, including the compounding areas, to be controlled to at least 95% efficiency. Mechanisms therefore had to be in place to reduce the hazardous air emissions by 95% from each major source of HAPs at the North Plant. At the adhesive tape plant, the efficiencies to be required by the new MACT standards were yet to be determined, and EPA was not expected to issue standards for adhesive tape coating for at least five years.

HAP emissions at 3M–Hutchinson in general were a subset of VOCs. The magnetic tape operation used a mixture of VOCs composed primarily of MEK and toluene, both of which were regulated HAPs, with the toluene regarded as the more toxic of the two. Lesser amounts of other VOCs that were not HAPs also were used, but 83% of the VOCs emitted at the plant were classified as HAPs. The adhesive tape operations at the South Plant used primarily non-HAP VOCs (such as heptane and naphtha) and smaller amounts of HAPs (such as toluene and methanol). Consequently, only 23% of the VOCs emitted there were classified as HAPs.

Making Environmental Improvements: The North Plant

Despite the fact that most of the production equipment at the facility had no control requirements under the Clean Air Act, 3M had made substantial environmental improvements. After trying unsuccessfully to find a way to eliminate or substantially reduce the use of solvents in AV tape production, 3M decided that the most feasible option was solvent recovery. At the North Plant, it installed, at substantial cost, a solvent-recovery system that enclosed the six magnetic tape coaters to ensure very close to 100% capture of emitted solvents. The system employed carbon beds to absorb solvents and distillation tanks to separate these captured VOCs into reusable solvent components. An elaborate system of above-ground pipes and tanks captured and processed the solvents and sent them back for reuse in production.

This solvent-recovery system was the main reason for a rapid and dramatic decline of more than 93% in North Plant emissions from about 14,000 tons of VOCs in 1989 to 931 tons in 1995 (see Figure 4-1). *This reduction was accomplished at the same time that production output increased by nearly 70%.* The amount recycled grew from about 2,500 tons of Toxic Release Inventory chemicals in 1991 to more than 11,000 tons in 1992. By 1995, 3M–Hutchinson was recycling more than 15,500 tons of chemicals, and the solvent-recovery system operated at an annual overall efficiency of about 98%, in contrast to typical systems in the industry that achieved 95% efficiency (the new MACT standard) or less.

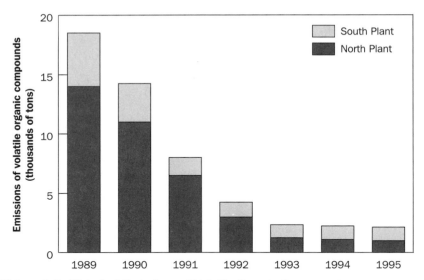

Figure 4-1. Historical Emissions of Volatile Organic Compounds, 3M–Hutchinson Plants, 1989–1995

Source: Minnesota's Toxic Release Inventory Data Base.

Because 3M had been required by law to control only two of its coaters, inclusion of the four unregulated coaters in the solvent-recovery system was a voluntary beyond-compliance initiative by the company. 3M did gain a regulatory advantage from the voluntary installation, however. By reducing HAP emissions 90% from 1987 levels before 1994, 3M was granted a six-year extension from the requirement to apply MACT to its emissions sources. Installation of the solvent-recovery system was also a good bottom-line business decision. It satisfied existing and future regulatory requirements while generating a positive return on investment. The system not only provided the North Plant with all the solvents it needed, but it also generated excess quantities that 3M sold to distributors.[6] In addition, it made sense for 3M to include all its AV coaters in anticipation of forthcoming MACT requirements.

Making Environmental Improvements: The South Plant

From 1989 to 1995, the adhesive tape operation in the South Plant reduced emissions by almost 68%, from 4,300 tons of VOCs to about 1,400 tons as production increased at the same time by about 70% (see Figure 4-1). There was less improvement at the South Plant because it was not possible to design a solvent-recovery system for the adhesive tape coaters. 3M made emissions reductions by means of source reduction and end-of-the-pipe control with thermal oxidizers, which were expensive add-on devices to burn off solvents in hot ovens. The use of thermal oxidation provided no direct economic benefit; its only justification was to meet 3M's internal goals of a 90% reduction in emissions by 2000.

Although the company did more than required by law, there remained significant room for improvement. Production at the South Plant was growing, and while emissions per unit of product were lower, actual emissions from adhesive tape operations did not decline significantly after 1991. 3M had been able to substitute water-based solvents for VOC solvents used in the production of some adhesive tape products and was working to expand the use of water-based systems. Fugitive (nonstack) emissions from VOC-based coaters could be reduced by completely enclosing the production line. In addition, for the production of some adhesive tape products, 3M was developing a new "hot melt" technology that significantly reduced VOC emissions at their source.

From 1989 to 1995, 3M had lowered VOC emissions for the North and South plants combined from 18,453 to 2,307 tons per year while at the same time production increased by about 70%. Most of these reductions had taken place by 1993 (see Figure 4-1), though total output at Hutchinson continued to increase by more than 17% from 1993 through 1995. Given that 3M intended to further increase production at Hutchinson, the extent to which it could maintain or enhance emissions reductions in the future would depend largely on the rapidity with which new less-polluting technologies could be developed.

Economic Difficulties

The environmental staff at 3M–Hutchinson was very supportive of the Project XL pilot project because they could see its potential benefits as they shifted their time from detailed compliance reporting tasks to more innovative pollution prevention and other environmental management activities.[7] The plant manager, however, was concerned about preserving jobs and having a predictable regulatory process in place so that the facility could convert expeditiously to new products in response to changing business conditions. Top 3M managers, who had set ambitious corporate environmental goals, sanctioned the 3M–Hutchinson XL experiment—provided it did not delay the timetable for product changes.

Indeed, computer diskettes and AV tapes had been a drag on 3M's earnings for several years, and significant changes were in the works. In contrast to adhesive tape products, the company had little proprietary advantage over its competitors in these markets. 3M could not compete with the manufacturing efficiencies of Asian companies, particularly in such products as diskettes and AV tapes that had very low margins. The outlook was that the competition was only likely to grow and that the pressure on prices would continue to increase. Excess worldwide capacity took away the incentive to remain in the diskette and AV tape businesses, even though sales of these products equaled about a fifth of the company's revenues in 1994.

In response to these factors, 3M, which was experiencing economic difficulties at the end of 1995, initiated a restructuring, the largest in its history. On November 14, the company announced that it would spin off its computer diskette and AV tape manufacturing businesses into a new company, which it called Imation. Close to 10,000 employees would move to the new company. Another 1,000 employees would be laid off.

Though the spin-off included 3M's magnetic tape business, the company retained the Hutchinson North Plant as part of 3M. 3M planned to convert the North Plant to the production of adhesive tape products. As a start in the conversion process, two of the plant's magnetic tape production lines were replaced with adhesive tape lines in 1996. At the request of several of 3M's large customers (for example, Disney), the company agreed to continue to produce large, high-quality visual recording master tapes for the next few years at the remaining magnetic tape production lines. 3M made the decision to phase out AV tape manufacturing at Hutchinson slowly because it wanted to retain as much of the workforce as possible. The company promised the community that management would eliminate as few jobs as it could.

The restructuring announcement came as a surprise to most 3M staff and employees and to the participants in Project XL–Minnesota. These changes, which greatly affected the Hutchinson facility, took place right before the design phase of the XL permit, creating significant uncertainty. This uncertainty further complicated the effort to reach an agreement.

For example, 3M was looking for ways to reduce Hutchinson's emissions by an additional 10% by 2000. Doing so would be difficult, especially now that 3M planned to phase out AV tape production. As long as one AV coater was left at Hutchinson, 3M would continue to run the solvent-recovery system, but it would have no more use for this system, no matter how efficient it was, once the AV coaters were totally shut down. Instead, a range of new environmental management strategies would have to be employed, and other ways would have to be found to reduce emissions from the new adhesive tape coaters that would be installed at the North Plant. Nevertheless, because of the company's ongoing research efforts, 3M's managers and engineers were confident they could bring new adhesive tape products to the market in an environmentally sound way.

The prospects of having a flexible XL permit for the Hutchinson facility became even more attractive to 3M because of the added uncertainty about future production at the North Plant. The new adhesive products 3M planned to introduce, whatever they were, would have specialized uses and command higher margins. Under the kind of permit the company envisioned, these new products and production lines would be introduced with a minimum of regulatory delay, an important advantage in the competitive setting in which 3M was operating.

Legal Uncertainties

In addition to the complexities discussed so far, a troubling issue was whether Project XL–Minnesota could be carried out without violating existing environmental laws and requirements. The project by no means was operating with a clean slate, because it had to contend with the massive structure of environmental laws and regulations created since the birth of EPA in 1970. If the statutory foundation for the pilot was insufficient, how could EPA proceed? The agency's view was that there was not much leeway in the law, and it was to the law that staff often appealed to justify their positions. Without additional legal authority, it was unclear if the agency could make the changes that XL required.

Many EPA staff participating in Project XL believed that the 3M–Hutchinson project stood on weak legal ground. It was unclear, for instance, if EPA had the authority to grant 3M facilitywide air pollutant emissions standards, waiver of individual emissions source permitting requirements, and reduced reporting of compliance. In comparison with congressional enactments, the Clinton–Gore declaration of XL policy did not have the legal standing to permit exceptions to the law (Clinton and Gore 1995). Some EPA staff members argued that without fresh enabling legislation, the legal means that the agency had at its disposal to experiment under Project XL were limited. In Minnesota, the state legislature had passed an Environmental Regulatory Innovations Law, which gave MPCA broad authority to experiment. But at the national level, Congress had not passed a similar bill. EPA's leaders were unwilling to expend their political capital to promote passage of such a bill, especially because the prospects for success, in the view of these leaders, were poor.

Site-Specific Rulemaking or Regulatory Discretion

The legal instruments that EPA's lawyers determined that the agency could use to issue a Project XL permit were site-specific rulemaking and enforcement discretion. Under site-specific rulemaking, many legal experts believed that leeway did exist in the law, especially in the Clean Air Act (Hirsch

1998). Companies were not compelled to follow end-of-pipe technology standards but could offer alternatives if these alternatives could show equivalent results.

Thus, the formal site-specific rulemaking provision in laws such as the Clean Air Act was viewed as a possible legal basis for Project XL. The other important element of leeway in the law, especially the Clean Air Act, was the discretion afforded to the EPA in dealing with a company technically in violation of the law. Under these circumstances, EPA generally worked with companies to get them back in compliance in a timely fashion. The problem with enforcement discretion as a basis for XL projects was that environmental groups or other third parties could go to the courts to challenge the plans that EPA had set in place. Site-specific rulemaking could be challenged, but not as easily, and it did not involve the stigma of being labeled a "violator." Because of this stigma, companies were uncomfortable using EPA's discretionary authority as a basis for XL agreements. However, applying site-specific rulemaking in such a broad way as ultimately was envisioned in the Hutchinson pilot was a novel approach; there was little precedent in prior law to back it up.

There was concern that environmental groups would sue EPA if an XL permit created significant exceptions to existing laws and regulations. EPA wanted to avoid a court challenge, if it could, but it recognized from experience that a significant percentage of its rulemaking ultimately was challenged in the courts. EPA lawyers felt that laws like the Clean Air Act did provide some flexibility for site-specific rulemaking, but they fully expected that even with this flexibility, there would be a court challenge.

EPA's regional staff was particularly concerned about the legal mechanism that would be used to authorize the XL permit. 3M had discussed the issue with EPA Region 5 lawyers, and the region's lead attorney on the project believed that the ideas 3M was pursuing raised issues that could not be easily resolved under the Clean Air Act.[8] If the 3M project was to proceed, EPA would have to invoke enforcement discretion for at least some elements in the permit. However, 3M and MPCA were seeking a site-specific permit, not enforcement discretion. The region's attorney consulted lawyers in the Office of General Counsel (OGC) at EPA headquarters, who did not agree that this issue had to be resolved immediately. They said it could be addressed later in the process after substantive issues about the nature of the project were better defined. OGC lawyers argued that, without knowing exactly what these substantive issues were, they could not determine which legal mechanism would be most appropriate or even if an acceptable legal mechanism would be available at all.

Some Region 5 lawyers also were concerned about another legal issue, one that related to the equal protection clause of the Fourteenth Amendment to the Constitution.[9] They were troubled that a precedent might be set wherein the regulatory flexibility granted to 3M in Minnesota would have to be granted to any company operating under similar circumstances in any U.S. jurisdiction. Environmentalists as well argued that under federal environmental statutes, EPA had to uphold uniform national policies. Their concern was that, without uniform national policies, states would compete with each other on the basis of the laxness of their environmental laws to attract industry. Environmentalists believed that great progress in centralizing authority had been made with the original passage of the clean air and clean water laws in the early 1970s. They did not want to see this authority undermined by Project XL. They believed XL could initiate a policy change of considerable magnitude without adequate review.

3M's View

3M also was worried about the prospects of a legal challenge. The company was less concerned about how an XL experiment would stand up in the courts than it was about avoiding a troublesome legal entanglement. It did not want to fight out the denouement of its XL proposal in the courts. Early on, therefore, before 3M's XL proposal had been accepted, in August 1995, it sent a letter to EPA's associate general counsel that summarized the legal authority it thought was available to carry out a pilot.[10]

In the letter, 3M praised the federal government for the dramatic advances it had made in pollution reduction, but argued that these advances should be consolidated and enhanced through simplified and cost-effective rules. According to 3M, this objective could, in large measure, be accomplished under existing statutory authority that granted EPA and the states the right to exercise flexibility. Under the Clean Air Act, for instance, pollution prevention was for the most part a state responsibility. This law, in 3M's view, relied on the general obligation that each party had to satisfy performance goals. It gave considerable latitude to the states to waive federal requirements to encourage innovative systems of continuous reduction, as long as there was no unreasonable risk to public health, welfare, or safety.

If the states achieved equivalent or better overall emissions reductions, 3M believed that they had the right to use regulatory mechanisms different from those commonly used by the federal government. In its letter to EPA, 3M emphasized that Project XL pilots would *not* be operating in violation of federal law, and permit applicants should not have to rely on consent decrees or any form of enforcement discretion that implied companies were in violation. The Clean Air Act did allow alternative approaches, if equivalent results were assured. The problem was to define equivalent results and determine if they actually were achieved.

Delegation of Authority to the State

Another issue that plagued Project XL–Minnesota was the extent to which a federal agency could delegate authority to a state agency. How much freedom to experiment could EPA give MPCA? EPA did not want to give up its role as the controlling body in setting national environmental policy. EPA was adamant that with regard to new initiatives, the agency would set policy *and* play a major role in the policy's development—at least in the early stages until difficulties were identified and eliminated. Although EPA insisted that this was its role, MPCA never completely accepted it. EPA remained steadfast that it could not grant a state agency the authority to create significant regulatory exceptions, out of concern that doing so would dilute the power of the national agency and undo the integrity of national law.

MPCA rejected the first draft of a memorandum of understanding (MOU) from EPA Region 5 staff concerning the authority the national agency had delegated to the state.[11] MPCA tried to frame an MOU on its own. In February 1996, it met with regional officials to discuss a draft of its MOU. Although the regional EPA officials agreed to revise the draft, they had no deadline to reach an agreement and did not remain in regular contact with MPCA. The two parties tried to work together to develop an MOU that would govern their relations, but MPCA and Region 5 did not agree about the extent of autonomy MPCA would have to design XL projects. The state agency wanted greater freedom of action than Region 5 thought appropriate (MPCA 1996e), and discussions were suspended without an MOU.

Region 5 staff members were disappointed by how MPCA and 3M briefed them about the proposed 3M–Hutchinson pilot.[12] They felt they received only general information about 3M and its facilities instead of a detailed discussion about what 3M intended to do. There was conversation about the broad structure for XL, and regional personnel accepted that 3M only would come forward with the particulars about execution later. 3M said it would be working with state government to develop XL proposals and would get back to the region only after the discussions were complete. Regional staff, though they encouraged 3M to work on its own, were frustrated with this outcome. They recognized that the Minnesota case was unique because MPCA had been selected as an XL participant, but they still believed that Region 5 should be a lead player. The region decided to give MPCA leeway at the beginning, but it reserved the right to step in at the end to determine if MPCA's proposal would be acceptable.

In the absence of clear legal authority, the informal relationships among the participants became of paramount importance. Without a well-defined legal mechanism or structure to guide them, the parties had to depend on teamwork and trust. The question was: Could the players in Project XL make up for the weak legal authority they had by trusting one another and collaborating effectively? (For discussion of this issue, see Sabel 1991; Hosmer 1995; Mayer and Davis 1995; Ruckelshaus 1996; and Weber 1998.)

Notes

[1]This chapter mainly relies on interviews with the participants and a site visit to 3M's Hutchinson, Minnesota, plant that took place on April 22, 1996. See the Acknowledgements for a description of the interviews.

[2]See MPCA 1996b and Abt Associates 1996.

[3]The predominant VOCs used were MEK, toluene, and cyclohexanone in the North Plant and heptane, naphtha, toluene, and methanol in the South Plant.

[4]There was a seventh uncontrolled coater in the North Plant that produced adhesive tape products but was only used infrequently.

[5]The "grandfathering–in" provisions of this act are very specific. The designation "new" facilities applies only to certain programs under Section 111 of the New Source Performance Standards. It does not apply to the entire act or an entire facility. Moreover, this exemption does not apply to "modifications" (that is, physical or operational changes to a facility).

[6]Solvents embedded in the solid substrate were extracted and captured during the production process.

[7]1996 interviews with 3M Company staff. See the Acknowledgements for a description of these interviews.

[8]Letter from M. Nash, senior counsel at 3M, to B. Grant, EPA Office of the General Counsel, in response to questions raised at April 15 meeting describing 3M–Hutchinson Project XL, April 18, 1996.

[9]Based on 1996–1997 interviews with EPA staff. See the Acknowledgements for a description of these interviews.

[10]"Statutory Authority for Project XL," letter from T. Zosel, manager of environmental initiatives at 3M, to L. Friedman, associate general counsel at EPA, with attachment, "Suggested Preamble Excerpts for Inclusion in a U.S. EPA Final Rule Governing Project XL Permit Program and Permit Authority," August 28, 1995.

[11]Ibid.; MPCA 1995c, 1996d.

[12]1996 interviews with EPA staff.

5

Drafting the
3M Proposal

The technical, economic, and legal challenges discussed in the last chapter were obstacles to a cooperative solution. They created uncertainty that made it more difficult for the parties to proceed. Technically, how could additional pollution reduction be achieved at 3M Company's Hutchinson, Minnesota, plants when so much progress already had been made? Economically, what was the future of the facility after 3M decided to spin off its audio and visual tape-making businesses? Legally, did the parties have the authority to draft an alternative permit? Given the history of their relations, could they work together to overcome these difficulties?

Despite these issues, the parties tried to draft an agreement, but doing so posed new challenges. How was the agreement to be developed and who would be involved? 3M took the lead. It shared its proposed agreement with the Minnesota Pollution Control Agency (MPCA). MPCA and 3M modified it and then shared the results with the stakeholder group, the Pilot Project Committee (PPC). The Minnesota participants believed that they had to be comfortable with the approach they were developing before they could show it to the U.S. Environmental Protection Agency (EPA). Nevertheless, achieving an agreement among the Minnesota participants was not easy. This chapter describes how, starting with 3M's initial draft, the company, MPCA, and PPC were able to develop a Project XL proposal they could support. In the next chapter, we show what happened when they presented this proposal to the EPA.

3M's Covenant

3M called its original proposal a "covenant" (3M Company 1996a). The covenant—drafted very much in the spirit of the beyond-compliance concept (see Chapter 3) that the company had proposed earlier—was consistent with what 3M believed to be the principles of Project XL. 3M's covenant set a standard for environmental performance higher than what the regulations required but let the facility decide on the best way to meet this standard.

From June 1995 to January 1996, 3M's XL team, led by Tom Zosel, worked to draft the covenant. The other members of the team were Dave Wefring, environmental regulatory specialist; Darryl Wegscheid, senior manager in the public affairs office; and Mike Nash, senior counsel. Experts in adhesive and magnetic tape operations from company headquarters and Hutchinson also participated. Their challenge was to collect the appropriate data and present them in a way that could be easily understood. The team conducted research and made judgements about what was environmentally responsible and technically feasible and about what would fit with existing and anticipated market requirements.

3M's covenant emphasized the company's values, goals, and past history with pollution prevention, and it focused on the enhancements the company was starting to make in its environmental management system (EMS). 3M's staff wanted to negotiate an agreement that made sense in terms of what the actual environmental protection requirements were for the Hutchinson facility, while eliminating unnecessary paperwork that did not have an environmental benefit. They wanted a permit that protected the public health, moved away from adversarial relations with EPA, and was as simple as possible. To protect the investment they made, they wanted the permit to last at least 10 years. Their draft agreement had three main features: (1) air emissions caps for volatile organic compounds (VOCs) and hazardous air pollutants (HAPs), (2) a waste ratio to serve as a simple indicator of progress in environmental performance at the facility, and (3) an advanced EMS.

Air Pollutant Emissions Caps

Given the large voluntary reductions in VOC air emissions already made at 3M–Hutchinson (see Chapter 4), the staff had difficulty arriving at suitable facilitywide limits (caps) on VOC and HAP emissions for a 10-year period. Although existing permits allowed the facility to emit close to 34,000 tons per year (tpy) of VOCs when operating at full production, the VOC cap would have to be much lower than the current allowed level to be in compliance with anticipated regulatory requirements and to clearly demonstrate superior environmental performance (SEP).

Current allowed emission levels were as high as they were because nine of the plant's 12 coaters, as well as other sources such as compounding or mixing tanks, came under the "grandfather clause" of the Clean Air Act and therefore were unregulated. However, EPA was expected to impose maximum achievable control technology (MACT) standards controlling HAP emissions from the magnetic tape (audio and visual, or AV) production facility. In setting a 95% emissions control efficiency standard, EPA would reduce allowed emissions at the Hutchinson facility to about 13,000 tpy of VOCs.[1] If 3M were to shut down an AV line and replace it with a new adhesive tape coater, the new production line would come under more stringent Clean Air Act requirements. Furthermore, most, if not all, of the four unregulated adhesive tape production lines in the South Plant were also likely to lose their grandfathered status during the next 10 years, either because of the imposition of MACT emissions restrictions or because of major modifications that would cause them to lose their exempt status and become regulated sources.

3M environmental managers expected that under the XL emissions caps, the facility would be free to make any modifications it saw fit without being subject to a permit review. This expectation meant that the company had to set the cap low enough to accommodate the potential regulation of most sources during the life of the permit. This determination was necessary to maintain the beyond-compliance philosophy that 3M claimed had inspired its XL proposal. Yet the timetable for shutting down existing AV coaters or converting them to adhesive tape products was not clear. Furthermore, it was not possible to predict which sources would undergo major changes or when these changes would take place because the answer to these questions would depend on future business conditions. 3M wanted the flexibility to make production decisions in response to business conditions and to take advantage of technological innovation and other opportunities. Indeed, this type of flexibility was one of the primary reasons that 3M was seeking a long-term XL permit.

In addition, the cap had to be set high enough to accommodate increased production at the facility. In 1995, the facility was operating well below full capacity, and 3M hoped to increase production substantially during the 10 years that the XL permit would be in force. If 3M set the caps too low, its conversion plans for the Hutchinson plants might be impeded because it could not meet the strict environmental requirements it had voluntarily imposed. Moreover, several new replacement lines that 3M was considering introducing at Hutchinson involved new and not entirely proven technologies. 3M engineers thought they needed an operating cushion for the potential emissions from these replacement lines. This cushion would allow them to work out any problems encountered with the new equipment. Thus, to arrive at reasonable limits, they needed a model they could use to estimate what 3M's emissions would be in the future. This was problematic, however, because of the many uncertainties about future operations at the North Plant.

Given these uncertainties, 3M environmental managers decided to assume that there would be no changes in production lines and that production would keep on increasing at the same rate it had been during the past few years. This meant that in about 10 years the facility would be operating at full production, which would be about 80% above current levels. This assumption provided as reasonable an estimate as any of the level of VOC emissions that might be expected, no matter what changes took place at Hutchinson.

In calculating the proposed VOC emissions cap, 3M also assumed that all of its tape coaters eventually would come under New Source Performance Standards regulation: The adhesive tape coaters would have to be controlled at 90% efficiency, and the AV coaters at 95% efficiency. 3M left its other sources, such as the mixing tanks, unregulated. These calculations indicated that total VOC emissions at 3M–Hutchinson should be capped at a reasonable amount below 5,300 tpy. Consequently, the company decided to propose a VOC cap of 4,500 tpy.

Because virtually all the HAPs emitted at the Hutchinson facility were VOCs, the VOC cap also provided an effective cap on HAP emissions. In the covenant, 3M did not propose a specific number for HAPs. Although the Clean Air Act required EPA to regulate all major sources of HAPs, the agency was still in the process of determining the MACT for industrial processes, which would determine how HAPs were to be regulated.

Because 3M anticipated production increases at Hutchinson but was uncertain of their nature and magnitude, it considered a cap set above actual emissions but far below allowable emissions to be challenging yet achievable. It would drive the company to make improvements, but at the same time give the company flexibility to deal with future production changes. 3M had been voluntarily reducing its emissions per unit of production for seven years, and it expected reduction to continue. But 3M could not guarantee this would happen—or that, if it did, these reductions per unit of production would be large enough to prevent total HAP and VOC emissions from increasing with rising production.[2]

A sensible solution might have been to introduce—instead of total emissions limits—caps on emissions per unit of production. This was not seriously considered, however, in part because of the complexities involved in defining units of production for the wide range of products to be produced at Hutchinson (some of which had not yet been invented). 3M was striving for as simple a regulatory framework as possible. The company also sought to retain at least some of the operating flexibility it had already achieved by the large voluntary reductions in VOC and HAP emissions it had made at the facility.

Furthermore, 3M staff believed that, although the emissions cap was set above actual emissions, it was set far enough below allowable emissions to give EPA, MPCA, and the public assurance that certain levels would not be exceeded. Allowed emissions would decrease over time but were not likely to approach the cap until the end of the 10-year life of the permit. The covenant proposed that any future environmental laws or regulations would not be applied to Hutchinson while the covenant was in effect unless MPCA or EPA demonstrated that the change was based on new scientific information and that failure to comply with the new law or regulation by the Hutchinson facility would result in material endangerment to human health or material harm to the environment. The covenant also proposed that upon 3M's request, when new operations were added to the Hutchinson facility, MPCA would increase the cap, albeit by an amount demonstrated by 3M to be beyond compliance with the then-existing regulatory requirements.

3M was reluctant to give up the large gap between the facility's allowable emissions and its current actual emissions. The company wanted to set the cap low enough to meet its beyond-compliance goal but high enough to enable it to retain some of the emissions cushion it was giving up by effectively regulating hitherto unregulated sources. Although 3M managers acknowledged that the cap was almost double actual emissions in 1995, in their opinion it was only a worst-case backup to reassure regulators and the public. They would do better than just stay below the cap. In their view, 3M's strong record demonstrated both that it could achieve SEP and that EPA and MPCA should be willing to rely on this record and trust the company.

The Waste Ratio

The covenant that the 3M staff proposed to MPCA was supposed to replace and consolidate into a single document the more than 20 permits currently required for the Hutchinson facility (3M Company 1996a). In addition to the emissions caps, it had a number of important features intended to help define SEP. One of these features was 3M's proposal to report a waste ratio number to the public as a way of informing people about the progress it was making with pollution prevention. The numerator of this ratio would be waste (W), and the denominator would be the sum of intended product (P), recycled by-products (B) that were productively reused and/or sold, and the useless remainder, or waste (W) that had no productive use $(W \div [P + B + W])$. The parameters W, P, and B were all reported by weight (pounds). Although experts agreed that reporting the ratio in terms of toxicity (i.e., harm-causing potential) would be better, limited scientific and technical knowledge made this problematic. A weight-based measure was easy to calculate and simple to use, and it would unite 3M employees behind the waste reduction effort.

The waste ratio was a representation of a facility's efficiency, and 3M's aim was to continuously reduce this number. 3M annually computed the waste ratio for each of its plants and reported the number to the responsible vice president. Each division vice president was given incentives, based partly on profit sharing and compensation, to improve performance in this area. In the five years since 3M had begun computing plant waste ratios, it had reduced the ratio for all of its facilities by an average of 32.5%, just short of its goal to improve by 35%.

Although 3M wanted to drive waste ratios at its facilities down as far as possible, the company made no explicit commitment in the covenant to do so. The covenant proposed only that the waste ratio for 3M–Hutchinson would be reported to the public; this number would be posted on 3M's website so the public could track its progress. The annual waste ratio for the facility was to replace the pollution prevention reporting requirements of the Minnesota Toxic Pollution Prevention Act, although 3M would still be subject to enforcement provisions.

Environmental Management System

3M regarded the EMS as a centerpiece of the covenant and another defining feature of SEP. The EMS was 3M's internal operating system for managing pollution at all its facilities. Ultimately, the EMS could be used to replace and consolidate many requirements and regulations. By linking together compliance and pollution prevention, it provided not only a framework for addressing environmental issues but also a mechanism for integrating environmental practices with the company's business operations.

The EMS had five elements:

1. regulations and 3M policies and standards,
2. environmental management plans,
3. environmental assessment,
4. environmental operating procedures, and
5. environmental goals.

The first element was designed so that a facility could identify which regulations and 3M policies applied to the facility. The second element would ensure that the plant staff were fully aware of all environmental issues at the facility and that they were organized to resolve any problems that arose as quickly as possible. The third element would promote self-assessment in addition to third-party audits. The fourth element would document actions and activities of the environmental managers at the facility. These managers were expected to write down the problems they had, the action plans they initiated, their schedules for solving these problems, and their assessment of the progress being made. They also were expected to integrate what they were doing with the facility's business plan. The final element of the EMS would formalize accountability for achieving 3M's environmental goals, which included, for 2000 (using 1990 baseline levels), a 90% reduction in all releases to the environment and a 50% reduction in waste. Pollution prevention approaches were given the highest priority. Ultimately, the EMS would provide facilities worldwide with an up-to-date regulatory information system that they could access easily (although, at the time 3M applied for its XL permit, not all the components of the EMS were operational).

3M's draft proposal identified its EMS as the mechanism with which to assure that the requirements of the covenant as well as other regulatory requirements were met and to assure that information on plant emissions was regularly provided to the community. The company proposed that its EMS would be audited annually and the results reported to stakeholders. This would verify that Hutchinson plant managers and engineers were adhering to the EMS plan for the facility.

Building Community Support

3M sought to build community support for the covenant and its definition of SEP in a number of ways. After the company realized that few people would notice the formal announcement in the local newspaper about the project, it started to consider additional ways to reach the public. The local stakeholder group, which was made up of Hutchinson residents, suggested that the plant manager and the MPCA XL project leader go on the radio to answer questions about the impact on the community. Company spokespersons appeared on television, and the company used public service announcements and articles on the editorial page of the local newspaper to provide information. 3M experimented with methods of making the formal and legalistic XL documents available to the public in jargon-free English.

The company carefully planned community information sessions. The sessions, promoted via cable television, radio, and the local newspaper, were held at the Hutchinson plant site, a community hospital, and a high school. The company displayed a series of posters with the story of Project XL and the company's history, philosophy, and record of environmental achievement. 3M staff members were stationed near the posters to answer questions. The company wanted the public to have accurate information so that when MPCA held a formal hearing, people would be informed enough to provide useful and intelligent feedback. 3M wanted to be certain that people in the community knew that it had made early voluntary reductions and that the company would not backslide with the new XL permit. 3M's stated intentions were to be as forthcoming as possible.

MPCA's Response

3M's goal was to work with MPCA to draft an agreement, but getting the project participants from 3M and MPCA to communicate and work together effectively was difficult at first. The number of people involved was unwieldy. The 3M facility was large and because the aim was to be multimedia, the only option was to have staff from different units and various media involved. Both sides found it hard to get away from the old regulatory paradigm of wait, react, and confront. Typically, 3M would initiate discussions without communicating its ideas beforehand to MPCA, and MPCA would react by asking questions about data or other aspects of a proposal. MPCA did have a comprehensive, holistic, multimedia list of actions it would have liked 3M to take, but it did not share the list with the company beforehand. The agency did not communicate its expectations in advance but waited instead for 3M to make the first move. Some progress was made, however, especially at an informal level in the communications among the technical people.

The logistics of gathering information and keeping stakeholders informed was complex, and the data requirements were greater than 3M had anticipated. From MPCA's perspective, 3M did not fully understand how much information had to be shared with the agency and with stakeholders if a deal were to be struck. Moreover, the company did not make decisions with the alacrity that MPCA would have liked. MPCA officials said that it took 3M from June 1995 to mid-January 1996 to formulate an initial plan (the draft covenant) for the Hutchinson facility—although, in fairness to the company, it was only named an XL participant in November 1995.

MPCA was having its own problems coordinating the work. Draft proposals floated around different divisions, and the staff had difficulty figuring out which one was most current, which comments came from which division, and how to reconcile them. MPCA had never done a multimedia permit of this scope, and it was uncertain how to proceed. No standard existed that the staff could use to evaluate environmental releases from a whole facility. The 3M project was time consuming, and it put pressure on XL project staff members, who had other duties to perform.

Setting the Caps: MPCA's Analysis

From MPCA's perspective, 3M's proposal had many shortcomings. The covenant, only 10 pages long, was a limited and sparse draft document that dealt mostly with air emissions. The approach was not conceived to be as much of a multimedia effort as some people at MPCA wanted. The covenant began by saying that 3M had a wonderful record in environmental protection and asked that it be trusted. 3M stated clearly that, given its strong corporate environmental policies, it should be entitled to a broad permit that was general in nature. In fact, EPA required a more complex proposal that consisted of two parts, an enforceable XL permit and a final project agreement that set out the goals of the project, explained why it fulfilled Project XL requirements, and provided background and details about the facility and the project.

In response, among the first actions that MPCA took was to limit the permit to tape and coating manufacturing and related support activities. If 3M switched operations, for example, to nonadhesive related medical products, those activities would be outside the scope of the agreement.

MPCA then began to take a serious look at the facilitywide VOC emissions cap of 4,500 tpy proposed in the covenant. For its part, MPCA faced a formidable and daunting task of figuring out what SEP meant and if this cap met the standard. According to EPA guidelines, environmental performance had to be better than it otherwise would have been under existing or anticipated regulations. Because actual emissions of VOCs at the facility were about 2,300 tpy, 3M's proposal allowed for total emissions to rise and therefore for a decline in absolute environmental conditions. In fact, as the proposal explained, production would go up, but there would be fewer emissions per unit of production; consequently, the environmental performance of the facility would have improved on a per-unit basis. Although MPCA considered setting the caps in emissions per unit of production, it received little encouragement from 3M or PPC. Normalizing emissions to production was difficult, given the large variety and uncertain variability in the number and kinds of products produced at Hutchinson. Thus, MPCA's XL team kept the flat emissions standard.

MPCA engineers examined 3M's air emissions model and recognized the great uncertainty involved in any estimate that could be used to establish a baseline from which to measure SEP. In an attempt to better define this baseline, MPCA decided that, for comparison, it would calculate a more-stringent lower limit for the total VOC emissions that regulations would allow when the facility would be operating at full capacity 10 years hence. MPCA engineers assumed that, at that time, best available control technology (BACT) standards would be applicable to all sources, which meant that 95% pollution control efficiencies would be required of all coaters. 3M had assumed 95% control levels for the magnetic tape coaters, but only 90% control levels for the adhesive tape coaters, in accord with current New Source Performance Standards. MPCA assumed instead that by 2006 there would be improvements in technology so that the adhesive tape coaters could achieve 95% efficiency.

With this assumption—and adopting 3M's unrealistic model of simply extrapolating the plants' current product mix into the future—MPCA engineers arrived at a figure for allowable VOC emissions that kept decreasing over time, reaching in 2006 about 3,300 tpy, a result well below 3M's esti-

Table 5-1. Setting Emissions Caps at the 3M–Hutchinson Plant

Criterion in Setting Cap	Cap
Allowable emissions (full production), 1995	33,989 tons per year (tpy) of volatile organic compounds (VOCs)
Estimate of allowable emissions *after* MACT introduced at North Plant	About 13,000 tpy of VOCs
Improvements at Hutchinson after 1987 (voluntary controls)	87% fewer emissions of VOCs, while production rose by about 70%
Actual emissions, 1995	2,307 tons of VOCs; about 1,100 tons of hazardous air pollutants (HAPs)
3M *estimate of total allowed emissions* at full production, *assuming* that NSPS would apply to all lines in 2006	5,300 tpy of VOCs
MPCA estimate of *total allowed emissions* at full production, *assuming* that BACT (95% control) would apply to all lines in 2006	3,300 tpy of VOCs; 1,900 tpy of HAPs
Caps in the proposed permit	4,500 tpy of VOCs; 3,000 tpy of HAPs

Note: MACT = maximum achievable control technology; NSPS = New Source Performance Standards; MPCA = Minnesota Pollution Control Agency; BACT = best available control technology.

Source: 3M Company, draft permit application and final project agreement, including the technical attachments.

mate of 5,300 tpy (see Table 5-1). During most of the 10-year life of the permit, however, MPCA's more stringent model still projected allowed emissions that were substantially above the 4,500 tpy cap proposed by 3M in the permit. Allowed VOC emissions only would fall below this proposed cap at or near the very end of this period because MPCA staff assumed that the changes at Hutchinson that would bring all of the production lines there under strict BACT standards would occur gradually.

The MPCA staff concluded that over the life of the proposed XL permit, cumulative VOC emissions at the Hutchinson facility would be far less than the otherwise applicable regulations would have required. This would be true even under the assumption made by the agency that these requirements would become considerably more stringent for adhesive tape production than they currently were.

Nevertheless, MPCA remained uncomfortable with the large difference between a VOC cap of 4,500 tpy and actual 1995 emissions of only 2,300 tons. The agency wanted greater assurance from 3M that environmental performance at Hutchinson would be superior. To provide this assurance, 3M agreed to incorporate into the permit as *enforceable* some of the requirements from the company's corporate environmental policy. Most notably, 3M would install BACT on all sources at the facility that emitted more than 100 tpy of VOCs.[3] In addition, 3M would control all new sources that emitted more than 40 tpy of VOCs, although a level of control as efficient as BACT was not explicitly required. The procedure for determining what constituted BACT (which included pollution prevention measures when appropriate) would be detailed in the company's EMS and would be arrived at with the advice and consent of MPCA. This provision covered all of 3M's coaters because when uncontrolled they all were capable of emitting much more than 100 tpy of VOCs. Thus, 3M agreed voluntarily to give up the existing grandfathered exemptions from the Clean Air Act for nine major VOC sources at its Hutchinson facility.

Hazardous Air Pollutants

HAPs were another difficult issue. 3M's draft covenant did not provide a number for the cap on HAP emissions, but soon the company came up with a figure of 3,000 tpy. MPCA engineers, conversely, found with their model that by 2006 total allowed HAP emissions—a subset of the VOC emissions—would decrease to about 1,900 tpy, well below the 3,000 tpy cap proposed by 3M. The state agency's analysis would have been justification for lowering the proposed cap on HAP emissions, but this never became an issue during the discussions among MPCA, PPC, and 3M. This was probably because the 4,500 tpy limit on VOC emissions effectively provided a more stringent constraint on HAP emissions than did the 3,000 tpy cap.

MPCA data for calculating levels of harm for emissions were limited to six of 40 hazardous chemicals emitted at Hutchinson. 3M was uncomfortable with the way MPCA used these data—obtained from the Minnesota Department of Health—to calculate the standards. 3M was mostly concerned with long-term, chronic carcinogenic impact and less concerned with short-term, acute effects. MPCA wanted 3M to deal with the short-term, acute effects as well. Also, MPCA was not satisfied with 3M's approach to record keeping, reporting, and compliance and wanted daily record keeping, not just annual data.

Other Problems

Another area of disagreement concerned emissions of small particulate matter. Again, 3M wanted compliance to be on an annual basis. Allowing for fluctuations from day to day, 3M could be in violation of air pollution standards on some days. MPCA preferred that 3M comply on a day-to-day basis. However, 3M was able to convince MPCA that the monitoring equipment needed to obtain daily emission numbers for small particulate matter would be too expensive to use at Hutchinson.

For VOCs, however, both parties recognized the need for daily monitoring, because the 4,500 tpy cap was to apply to total emissions calculated each day for the previous 365 days. This requirement presented a short-term problem for 3M. The company had installed a continuous monitoring system for its sophisticated solvent-recovery unit at the North Plant, but although it had plans for doing the same for its adhesive tape production lines, it had not yet done so.

MPCA also had to reconcile the role pollution prevention (P2) concepts would play in Project XL. Initially, the agency was not prepared to have P2 play a large role, but P2 advocates within MPCA felt that the agency was not placing sufficient emphasis on P2 concepts and applications in XL. When agency staff made site visits to the facility and realized how much energy was being wasted by the pollution control equipment and how preventable amounts of other pollutants were being emitted, they better understood P2 and how it could be a means to achieve SEP. Nevertheless, most of the discussion at meetings involved air quality and, to a lesser extent, hazardous waste. Only a small fraction of time was spent on P2.

Reaching an Agreement between MPCA and 3M

After the draft covenant was reviewed by a wide spectrum of people in the agency, MPCA received 15 pages of detailed, technical comments. The agency then held a series of meetings with 3M to resolve these issues. Reaching an agreement about the proposal looked like a difficult task. Many people attended these meetings, and they often left with different impressions of what had taken place. Unfortunately, a system for recording agreements that had been reached was not instituted. Thus, some participants were concerned that people would come to subsequent meetings and reverse their earlier positions. Tom Zosel, 3M's manager of environmental initiatives and XL team leader, realized that the linear nature of the project design process was hampering progress.[4] 3M independently developed its covenant. It then sent the covenant to MPCA, which made comments and sent the covenant back to 3M. 3M then responded to the comments and sent the covenant back to MPCA—and so on. This linear process was not productive. Moreover, 3M and MPCA had long negotiated from opposing sides. They had not always worked cohesively toward common goals. Zosel believed that 3M staff and MPCA staff were engaging in posturing in their exchanges. The two sides needed a different way to relate. Zosel thought they should use a development process that was parallel in nature, not linear and sequential. He wanted the two sides to work together as a team.

At a meeting arranged by Zosel, he placed 3M and MPCA staff members randomly at a large round table. All participants were asked to say what their goals were, why they were involved, and what they believed the issues to be. These were written down on a board for everyone to see. According to Zosel, when 3M staff and MPCA staff carried out this exercise, agreement about certain basic points was nearly unanimous. Both sides agreed that the reason that they were involved in XL was to find a better way to regulate. This desire came out of frustration with costly, inefficient processes that did not lead to better environmental results and that hurt industry. Both sides also thought that better environmental protection was possible and that it could be achieved in a less complex, less bureaucratic way. Zosel thought that once MPCA staff and 3M staff understood that they agreed on these points, tension between them decreased. At this point, each side began to see issues from the perspective of the other side. They became committed to reaching an agreement that was environmentally responsible and flexible and met the goals of the agency, industry, and the community. Remaining issues became easier to resolve.

Subsequently, MPCA and 3M agreed to the 3,000 tpy cap on HAPs (MPCA 1996a). MPCA was confident that the two sides were moving in the right direction and was certain that 3M's emissions would be below regulatory requirements. MPCA was reassured by a feature in the XL permit that called for regulatory reanalysis at years 2, 5, and 9 of the 10-year agreement to determine compliance with existing laws. Neither the 4,500 tpy VOC cap nor the 3,000 tpy HAP cap was fixed once and for all; they could be revised to be more stringent to accommodate changes in the law or new scientific findings.

The Pilot Project Committee's Response

3M and MPCA staff then presented the proposed agreement to the PPC, MPCA's stakeholder advisory group and responded to many of the PPC's questions.[5] 3M provided some data directly to PPC (PPC 1995) and on April 22, 1996, allowed PPC members to go on a tour of the Hutchinson North and South plants.

Superior Environmental Performance

PPC, which consisted of environmental managers, consultants (both legal and scientific), environmentalists, and academics, argued that under XL, three criteria should be used to assess the environmental performance of the facility (PPC, notes from monthly meetings with MPCA, 1996–1997):

- actual declines in emissions from a 1995 base or 1993–1995 average,
- declines in emissions per unit of production, and
- significantly lower emissions than allowed under both current and future regulations.

PPC took the position that only after the XL pilot had run its course could the actual extent to which SEP had been achieved be determined. Guaranteeing SEP before the fact, according to PPC, was against the experimental nature of the pilot and was not a prerequisite for proceeding. The pilots were merely tests (Dorf and Sabel 1998), and by their nature they involved some indeterminacy. If a priori guarantees were imposed, XL pilots were not experiments but conventional regulation pursued by other means. After-the-fact evaluation, according to PPC, should be used to determine the extent to which a pilot achieved its promised SEP.

Another issue that came up in discussions with PPC was how BACT would be determined. 3M's corporate environmental policy guidelines were embedded in the proposed permit, which required 3M facilities to install individual controls on all new sources that released more than 40 tpy of VOCs and to install BACT on all sources (i.e., both new and old) emitting more than 100 tpy of VOCs. The definition of BACT would be worked out by mutual agreement between MPCA and 3M as part of 3M's EMS, thus giving the company operating flexibility.

Members of PPC were concerned that these source-by-source requirements would discourage pollution prevention approaches. The agreement to rely on BACT was meant to give MPCA, stakeholders, and the general public reassurance that 3M would not take advantage of the fact that the VOC and HAP emissions caps were set well above actual emissions levels. Although this part of the permit was a throwback to conventional regulation focusing on source-by-source controls, P2 could be used in the definition of BACT. The extent to which these source-by-source requirements might lower incentives for innovation would depend on the meaning given to BACT by negotiations among 3M, MPCA, and stakeholders. The permit placed BACT decisionmaking at this level, where the facility and its problems presumably were best known and where 3M and MPCA had a working relationship. If an agreed-on procedure for determining BACT could be developed among XL participants and incorporated into the company's EMS, 3M would have some latitude to select the form of BACT best suited to the problem at hand, provided it was consistent with the agreement and the EMS.

Positive Features

PPC members who represented Minnesota environmental groups had reservations but nevertheless were supportive of the agreement that was being forged. An especially important reason that they were satisfied was because of the pollution prevention measures in the proposal (Marcus, Geffen, Sexton, and Smith 1996b). They pointed to the following positive features.[6]

Multimedia Characteristics. Though focusing on air pollution, the permit was holistic, including air, water, and waste issues. The XL permit would replace 22 individual permits.

Measurement and Reporting Mechanisms to Encourage Pollution Prevention. Although there was no explicit requirement in the permit or final project agreement that 3M would adopt P2 approaches to achieve SEP, the monitoring and evaluation procedures that relied on public disclosure of mass balance input–output data encouraged them. In addition, 3M would have to reveal whether P2 measures were included in individual plant projects having an environmental impact. The proposed final project agreement committed the facility to report:

- corporate waste ratios;
- the planning and inception of projects that it expected would have an environmental impact, including whether source reduction or recycling efforts would be present, the project's effect on environmental releases of Toxics Release Inventory chemicals, or, if rejected or delayed, an explanation of why the project was not feasible;
- descriptions of how projects were enabled by Project XL (if possible, expressed in percentage terms as an enabling factor);
- the percentage reduction in pollution achieved; and
- when data were readily available, information on water use, energy use, solid waste generation, and product-to-package ratio.

The disclosure of these numbers was considered to be a very important incentive to carry out experiments in pollution prevention.

Mass Balance: Input–Output Analysis. The caps to which 3M agreed simplified the air emissions measurement process because plant engineers could apply input–output analysis to the entire facility instead of having to meticulously track inputs and outputs at every production unit. 3M recorded the materials that were the inputs to the different products made at Hutchinson and had accurate measures of the material contents of the outputs of the final products. These masses, according to the law of conservation, had to be in balance. The difference between the inputs and outputs was thus the waste produced, on which 3M promised detailed reporting. Some of the waste was processed to create useful by-product (e.g. the North Plant's solvent-recovery system produced more than enough solvent to run the AV tape coaters). The remaining waste (solid, liquid, and gaseous) was the quantity used to compute the waste ratio.

This was not the end of the story, however, because this waste had to be treated and/or disposed of and the amounts carefully tracked. The gaseous waste, overwhelmingly VOCs, was almost entirely captured and incinerated in large thermal oxidizers whose operating efficiencies had to be carefully monitored. Because each of these control devices processed air emissions from several production units, plantwide mass balance analysis was possible. VOC capture rates had to be measured as well. Plantwide mass balance analysis became a useful measurement tool, but monitoring air pollutant emissions at such a complex facility continued to be difficult.

An Innovative Environmental Management System

At the time the new permit was being designed, 3M was in the midst of a major upgrade of its EMS. The permit reinforced this effort and established the EMS as the key compliance element for the Hutchinson facility. A sophisticated EMS would not only ensure compliance with existing environmental regulations but also move the facility into more proactive, "design for the environment" modes that would integrate the environmental protection function into the overall strategic operations of the organization. The EMS, when completed, was to include a detailed section on goals that would commit 3M to:

- evaluate all opportunities for preventing or reducing pollution, using the waste management hierarchy so that source reduction would come before other alternatives such as recycling;
- establish site-specific waste ratio goals;
- set goals for projects expected to make reductions in Toxics Release Inventory chemicals;
- if feasible, establish energy and water use improvement goals, criteria and conventional water pollutant reduction goals, and solid waste reduction goals; and
- maximize product recyclability and minimize environmental impact following consumer product use.

Stakeholder Involvement. Under Minnesota law, MPCA could revoke the XL permit if 3M failed to satisfactorily address a substantive issue raised by a majority of the members of either its stakeholder group (PPC) or 3M's local stakeholder group.

Reporting Environmental Results—the Internet. After 3M developed a website for the Hutchinson XL project, the company would post and periodically update actual emissions data (see Orts and Murray 1997).

PPC Endorsement

After hearing from the 3M and MPCA on issues that were of concern to it, PPC endorsed the proposed permit. Though drafting the proposal was a formidable undertaking, at this stage the XL proposal had support from the Minnesota contingent. 3M, MPCA, and PPC—including its environmental organization representatives—felt comfortable with it. This obstacle—getting the Minnesota participants to agree—had been overcome. This proposal had been developed, however, without EPA Region 5, and significant problems lay ahead as 3M and MPCA prepared to present the proposal to EPA.

Notes

[1]The North Plant at Hutchinson had a solvent-recovery system that experienced efficiencies considerably higher than 95%—the new MACT standard. EPA considered that 95% was a more reasonable standard than the 97–99% efficiencies achieved at 3M's AV plant. As the North Plant's recovery system aged, it was not clear whether these extraordinarily high efficiencies could be maintained at a reasonable cost.

[2]If we simply extrapolate past performance forward 10 years, production would increase by 80% at the facility, and if air pollution per unit of production remained unchanged, Hutchinson would generate 80% more VOC emissions or 4,140 tpy. Perhaps, 3M chose a cap of 4,500 tpy as an initial bargaining position expecting to accept a lower figure later. Strangely enough, the issue—formulated in this way—was never discussed.

[3]To give an idea of the scale of VOC quantities involved, a typical coater at Hutchinson that uses organic solvents for tape production emits VOCs that, before control, are measured in thousands of tons per year.

[4]Based on 1996 interviews with 3M staff. See the Acknowledgements for a description of these interviews.

[5]Notes from a presentation to PPC of 3M's plans for XL permit at Hutchinson, 1996.

[6]Comments by Minnesota Center for Environmental Advocacy, Sierra Club Northstar Chapter, and Citizens for a Better Environment on the Proposed Minnesota XL Permit No. 96-01. Issued to 3M Company for its Hutchinson Tape Manufacturing Plant, June 28, 1996; also MPCA 1996a, 1996c.

6

Gaining EPA Support for the Agreement

The parties trying to create an environmental agreement must not only agree among themselves. They also need the support of other parties—who are often more influential, who may not have been involved in the original negotiations, but who nonetheless must give their approval if the project is to go forward. Without the support of these other parties, a project can come to a halt. Thus, an additional obstacle to cooperative environmental solutions is obtaining support from parties beyond those who formulate the initial agreement. Negotiations are not conducted and concluded only once; they are repeated again as more groups are brought into the process and an attempt is made to secure their support. These new groups bring to bear additional concerns, which are often quite legitimate. They may introduce entirely new objections that the original parties to the deal missed, suppressed, ignored, or did not consider important. Negotiating a cooperative environmental solution, therefore, is a not a one-time event, but an iterative process. An agreement emerges only if it can withstand the rigors of this process.

By the end of May 1996, the 3M Company, the Minnesota Pollution Control Agency (MPCA), and its stakeholder group (the Pilot Project Committee, or PPC) had reached a consensus on a Project XL proposal, and the next step was to secure the buy-in of the U.S. Environmental Protection Agency (EPA). As 3M and MPCA began discussions with the agency about the proposed XL permit and final project agreement (FPA), it became clear that several problems remained. In a dramatic meeting between the Minnesota contingent and officials of EPA Region 5, these issues came to a head. The new objections EPA voiced were difficult ones for the Minnesota contingent to handle. This chapter describes MPCA's efforts to gain support for the proposed agreement and its reactions once EPA showed opposition to the deal.

Problems in Relating to EPA

Although EPA had approved Minnesota's XL proposal in November 1995, MPCA was not clear if it could act independently to approve the 3M–Hutchinson deal (see Chapter 4). There was no memorandum of understanding to delegate the authority to manage XL projects from EPA to the state agency (U.S. EPA 1995a; MPCA 1995c, 1996d). If MPCA did not have the right to approve the project, what authorization from EPA was required before the project could proceed? Who at EPA would decide, using what process, and within what period?

MPCA Commissioner Charles Williams had met earlier with Jon Kessler, the director of regulatory reform at EPA, to discuss this issue. Although Kessler was vague about the extent of the authority EPA would or could delegate to MPCA, Williams and his staff proceeded under the assumption that they had been granted considerable authority simply by virtue of MPCA's selection to be a participant in Project XL.

MPCA, in addition, was not sure which part of EPA it should be dealing with to resolve XL-related questions.[1] Who at EPA would make final decisions on Project XL? Who were the main EPA participants? Should MPCA communicate with EPA headquarters, the regional office, the program offices, or all of these offices? MPCA staff sensed that if they could gain the support of the EPA regional office, then EPA headquarters would go along with the proposal, but MPCA was not sure who in the regional office was the decisionmaker, nor did they understand the complex interrelationships among the region and multiple EPA headquarters offices. During the period when the Minnesota participants were developing the 3M XL project, MPCA did maintain some contact with EPA staff by telephone. Conference calls between MPCA and EPA typically had involved a diverse collection of EPA staff from the Office of Policy, Planning, and Evaluation (OPPE), the Office of the General Counsel, other program offices at headquarters, and Region 5 offices. Discussions tended to be free-wheeling, protracted, and unsatisfying. Consequently, up to the time that they met face to face with the Minnesota group in May 1996, EPA participants did not have a clear idea of the details of the 3M project. Nor did MPCA get an unambiguous response from EPA as to what would be acceptable and what would not.

Different messages seemed to come from different groups at EPA. On the one hand, the staff of OPPE believed that just doing better than legal requirements was not superior environmental performance (SEP), especially for a company like 3M with a record of going well beyond compliance. OPPE staff members wanted more direct benefits for the environment than citizens would receive under the current system. Their definition of SEP involved achieving environmental results superior to those that otherwise would have been achieved under existing regulations.

On the other hand, some of Project XL staff members in EPA Region 5 were concerned that XL proposals did not test explicit hypotheses, but instead merely proposed trade-offs that would be helpful to the company. Where was the analysis of the problems with the existing regulatory system? How would success or failure shed light on these problems? According to some EPA Region 5 staff, no one at EPA or in industry understood or was receptive to the idea of Project XL as a test of hypotheses. Most of the proposed XL projects that EPA was reviewing focused on incentives for SEP: EPA would give the company a degree of operating flexibility in exchange for a proportionate improvement in environmental performance. The XL projects were not designed to test the hypothesis that companies would be more likely to achieve SEP if they had the regulatory flexibility to reach environmental performance goals in a more cost-effective manner.

3M participants as well as those from the MPCA had experienced frustration in their discussions with EPA. They complained that the agency had set up too many barriers by not providing adequate leadership and structure for the XL decisionmaking process. Early in the process, the 3M participants believed that Region 5's XL team was overly critical and unsympathetic to its proposal, and 3M contacted EPA headquarters to try to obtain a more favorable hearing.[2] When Region 5 staff learned of 3M's efforts, they were angry and resentful at what they viewed as an attempt to circumvent their authority. They worried that political pressure was being put on them to approve the 3M proposal and that EPA headquarters was siding with 3M.

For Region 5's XL team, the 3M–Hutchinson project meant an additional burden on top of their regular workload. The team members had devoted substantial time and effort to XL, even to the extent of neglecting some of their other duties. Although they were sympathetic to the reform ideas behind Project XL, the regional staff was trying to overcome obstacles to experimentation that they believed were imposed by the Clean Air Act. They were also exasperated with 3M and MPCA for not sharing data in a timely manner and for not including Region 5 in planning meetings and project negotiations.

These complications came to the forefront in March 1996 when the director of the EPA regional Air Quality Division ordered his staff not to participate further in the 3M–Hutchinson XL project. He wanted his staff to focus instead on reducing the backlog of air permits accumulating on their desks, and he ordered them to stop participating in XL conference calls and to provide no more technical assistance to MPCA. In comparison with the thousands of permits the region issued, he believed little harm would follow from approving a few experimental pilots. In his opinion, if MPCA wanted to pursue the XL project on its own, Region 5 should allow it to do so. He favored both approving the 3M–Hutchinson proposal, despite objections from Region 5 lawyers, and also letting future litigation take care of itself.[3]

In May 1996, the head of the Air Quality Division allowed his staff to resume its participation in the project when it became clear that the division's expertise was needed to analyze and interpret the 3M–Hutchinson proposal. Meanwhile, however, during the Region 5 hiatus, MPCA continued working with 3M and PPC to design an acceptable XL proposal, and there were limited contacts with other Region 5 staff—suggesting that MPCA may not have fully comprehended the seriousness of alienating the Air Quality Division's staff.

Presenting the Proposal to EPA Region 5

Thus, the MPCA staff was not able to gauge EPA's reaction to the 3M–Hutchinson XL pilot and was not sure what the agency would do. MPCA decided to move forward by presenting the EPA Region 5 office with a draft permit and FPA at a face-to-face meeting. The Region 5 project managers agreed with this approach, and the meeting was scheduled for May 20, 1996, in Chicago. Adding to the pressure already on MPCA was 3M's position that the company needed to have a decision on the XL proposal by June 1 because of anticipated process changes at the Hutchinson plant. Without resolution, 3M would have to apply for permits under the conventional regulatory system.

The role that the stakeholders would play in the upcoming Chicago meeting was unresolved. The MPCA staff thought that the stakeholders were an important part of the process and that their support at the meeting would be crucial. Therefore, the MPCA staff believed that some PPC members should attend the Chicago meeting. 3M, in contrast, was wary of having PPC representatives attend because the company was unsure what position they would take on various permit issues. After much discussion, 3M agreed that representatives from both PPC and the Hutchinson stakeholder group could attend the meeting.

Attending the Chicago meeting from EPA were 12 officials from various programs in Region 5, including staff from the offices of regional counsel, air quality, waste management, water enforcement, and Superfund. Chief spokesperson for EPA was Dave Ullrich, the deputy administrator for Region 5. Lisa Thorvig was the chief spokesperson for the Minnesota delegation. Among the attendees were Tom Zosel, 3M's manager of environmental initiatives and XL project leader; Jim Bauman, plant manager of 3M's Hutchinson plant; Mike Nash, senior counsel for 3M; Andy Ronchak, Project XL coordinator at MPCA; Anne Seha, assistant attorney general for the state of Minnesota; and representatives from PPC and 3M's stakeholder group.

Although EPA had had the full 3M–Hutchinson XL proposal for only about a week, MPCA expected the meeting to produce an EPA decision about whether it was acceptable or, if not, what changes were needed to make it acceptable. Just before the meeting, however, Region 5 announced that it was not prepared to make a decision. The region reserved the right to delay deciding for at least a week, until it had the chance to review carefully the documents submitted by MPCA and to coordinate its response with other EPA offices in Washington, D.C., and elsewhere.

The meeting began with a spirited exchange between Ullrich and Zosel. Ullrich noted that EPA appreciated 3M's willingness to step forward and be innovative, but he was uncertain to what extent the environment would be better off under the 3M–Hutchinson proposal. EPA's obligation to Congress and the public was to guarantee environmental protection, not economic performance. Ullrich made it clear that the determining factor in EPA's decision about the proposal would be 3M's ability to leap over this deep "chasm" by demonstrating, without qualification, that it could deliver a superior level of environmental improvement.

Zosel responded that the environment, the economy, and the community were all important considerations and that each would benefit from an agreement that provided both environmental benefits and lower costs. He made the point that in addition to the investment costs needed to achieve an environmental benefit, considerable transaction costs were also involved. When 3M had to make rapid product and process changes in response to global competitive challenges, it was often hamstrung by a long, drawn-out permitting process. The 3M–Hutchinson proposal allowed 3M to focus on overall performance goals at the plant, rather than to rely on source-by-source permitting, which could take six to 18 months per source.

Ullrich replied that though cleaner, cheaper, and smarter regulations were an appropriate goal that was supported by EPA, the fact remained that EPA's primary responsibility was to protect the environment and public health, irrespective of corporate convenience. He reiterated that EPA was accountable to U.S. citizens and their elected representatives, not to corporate America.

After this unscheduled and unplanned debate, Thorvig of MPCA began her presentation, summarizing reasons for MPCA's support for the 3M–Hutchinson proposal. Among these were the innovative features of the proposal, such as stakeholder involvement; a departure from command and control that would give 3M relief from time-consuming permit applications, thereby enabling the company to respond rapidly to changing market conditions; treating the facility as a single source; reducing total emissions of hazardous air pollutants (HAPs) and volatile organic compounds (VOCs); and the incorporation of worthwhile goals for environmental improvement, such as the use of advanced technologies to eliminate or greatly reduce the use of solvents in the company's tape manufacturing processes. In addition, Thorvig emphasized that the agreement provided for reviews in 1998, 2001, and 2005, at which points 3M had to show it was providing direct environmental benefits or else MPCA could terminate the project. Moreover, 3M had committed to enforceable emissions limits well below allowable levels, to a third-party audit of its environmental performance, and to limiting operation of its audiovisual tape coaters in the event that its solvent-recovery system encountered problems. Finally, Thorvig noted that the agreement had been negotiated with the active participation of the two stakeholder groups.

After representatives of both the Hutchinson stakeholder group and PPC spoke briefly in support of the agreement, Ullrich responded. He pointed out that the emission caps established in the permit, though below current allowable emissions, were greater than current actual emissions, which was cause for concern. He also called attention to the fact that the proposed Project XL permit would allow 3M, as long as it stayed within the permit caps, to avoid complying with Title 5 of the Clean Air Act, including New Source Performance Standards, application of maximum achievable control technology, and source-by-source reporting requirements. These rules were written into federal law, which had been passed by Congress and signed by the president. Ullrich reminded 3M that the company had a heavy burden to provide justification as to why EPA should waive these legal requirements.

Ullrich went on to talk about the advantages to 3M of the proposed agreement, emphasizing that anticipated cost savings for the company would not be available to its competitors. How could EPA justify providing such economic benefits to a single company, particularly when the environmental benefits were based on promises and possibilities rather than a legally binding commitment? The 3M–Hutchinson permit was likely to be challenged in court, and Ullrich was concerned that if environmental benefits were not guaranteed, EPA might lose a court fight. EPA, in Ullrich's view, could not justify granting an exception to the Clean Air Act that did not secure direct environmental advantages. If 3M failed to live up to these promises, the only remedy was public embarrassment—an inadequate mechanism.

Ullrich also mentioned the concerns of his legal staff that the proposed agreement would be precedent setting. Under the equal protection clause of the U.S. Constitution, EPA might have to offer similar permits to other qualified applicants.

3M's representatives responded that the Hutchinson plant was unique and that no precedents would be set. The company was planning higher levels of production, and the proposal called for higher levels of control, which meant that HAP emissions would be below prevention of significant deterioration requirements. It was 3M's intention that any new sources would be controlled with state-of-the art technologies, and the company was testing new approaches that eliminated solvents from the tape manufacturing process. However, it was not yet clear whether the new technologies would be successful. 3M argued that EPA should be sensitive to the reality that technological progress depended on obtaining the necessary regulatory flexibility. In 3M's opinion, EPA should encourage technological innovations by environmentally responsible companies, not thwart their efforts by implementing rigid and unrealistic regulations, which locked firms into outmoded paths that benefited neither the environment nor the economy.

EPA representatives countered that 3M was seeking credit for good deeds it had carried out in the past and that this was not consistent with either the Clean Air Act or the founding principles of Project XL. They believed XL was meant to be forward looking and to obtain guarantees of future environmental improvements in exchange for flexibility. It was not a reward for exemplary corporate behavior that occurred prior to XL.

Some EPA representatives expressed the view that, *under current law*, 3M was likely to build on its past successes and achieve future environmental breakthroughs anyway. The company had previously achieved voluntary environmental improvements despite the constraints of the existing regulatory system, and there was no reason to believe that it would not do so in the future. Why should the agency take the risk of overriding portions of the Clean Air Act to secure accomplishments that 3M would probably make anyway?

Without guarantees of additional environmental benefits, EPA saw no reason to approve the agreement. EPA did, however, tell MPCA it could put the proposed agreement out for public comment, but the final approval would depend on further EPA review by regional and headquarters offices. The MPCA representative indicated that the state would proceed with the public comment process despite the absence of EPA approval. On May 23, 1996, the MPCA decided to place the proposal on Public Notice.

Presenting the Proposal to EPA Headquarters

During the public comment period, the Minnesota contingent (MPCA, 3M, and stakeholders) met again with EPA in Chicago; this time, staff from EPA headquarters offices, including the air quality program in North Carolina, was also in attendance. Interviews and notes taken by several of the participants form the basis of what we know about this meeting.[4] This was the first face-to-face meeting between EPA headquarters staff and MPCA. At the meeting, MPCA argued that based on an average of 3M–Hutchinson's actual emissions during the preceding five years, the proposed emission caps in the XL permit constituted SEP.[5] The MPCA spokesperson pointed out that if 3M were not operating its voluntary controls, HAP emissions at Hutchinson would be 12,000 tons per year (tpy) and VOC emissions would be 17,000 tpy, still far below the 34,000 tpy permitted by law.[6]

For its part, EPA expressed concerns about the issue of "backsliding." What if, during the time that the proposed permit was in place, 3M's emissions, even adjusted for production levels, were to go up? 3M answered that this scenario was not likely to take place because the company was not going to turn off its voluntary controls. The caps on VOC and HAP emissions were in the permit to prevent 3M from doing a complete turnaround and dramatically increasing actual emissions. 3M's position was that if it failed to meet the caps, then the XL experiment was over. The company believed it had a reasonable chance of meeting or exceeding the ambitious goals in the FPA. 3M was committed to making a good-faith effort.

EPA was unmoved by these arguments and asked for additional specific and enforceable performance criteria. EPA also proposed that either party (EPA or 3M) should have the right to immediately terminate the agreement if the other party was not living up to its commitments. 3M maintained that the right to termination already existed under site-specific rulemaking, whereas EPA argued that it would take several months under due process to end the agreement. EPA held that it should have the authority to revoke the agreement unilaterally without publishing its reasons in the *Federal Register* and waiting for public comment.

EPA also raised questions about the scenario wherein 3M's emissions per unit of production increased, but total emissions stayed under the caps for VOCs and HAPs. EPA wondered if the agency would have sufficient grounds to terminate the agreement if the caps were not being violated, but only because production was down. EPA also wanted to know what would happen if 3M was not forthcoming with information. For instance, according to the FPA, 3M was supposed to report the pollution prevention projects it conceived, accepted, and rejected, including the reasons for its decisions. What recourse would EPA have if 3M did not provide this information?

EPA wanted stipulations in the agreement that would prevent, or at least discourage, these types of situations from arising in the first place. The agency also wanted language in the permit that specified 3M's remedial actions in the event that these problems did arise. 3M objected to these additions, arguing that it deserved EPA's trust and that the agreement must articulate SEP goals, not mandate enforceable limits that put 3M at risk of being in noncompliance.

The Aftermath

Despite these conflicts, the MPCA staff believed that this second meeting with EPA had gone relatively well. They felt that EPA's objections were not so much to the 3M–Hutchinson agreement per se, but rather focused on the broader issue of how to accommodate and reconcile Project XL with the Clean Air Act. MPCA was concerned about how bureaucratic EPA was being and with the fact that different factions and fiefdoms in the agency did not seem to be talking to one another. Some MPCA staff members questioned the extent to which EPA had accepted the charge from the Clinton administration to reinvent environmental regulation.[7] Nevertheless, MPCA still hoped that the additions, clarifications, and refinements sought by EPA could be accommodated with only marginal changes to the permit and FPA.

At about this time, 3M began to have doubts about the agreement.[8] The company was coming to grips with the fact that it was paying a high price in transaction costs for its efforts to reach an XL agreement, which might never come to pass. 3M's immediate problem was the need to obtain an interim permit to cover the changes it was making at the Hutchinson facility, and the company was worried that the XL negotiations would either drag on for too long or end in failure. 3M began to consider reverting to conventional permitting.

At this critical juncture, Commissioner Williams left MPCA and was replaced by Peder Larson. Larson was a former congressional staff person and had been MPCA deputy commissioner under Williams. Lisa Thorvig became deputy commissioner, and for the time being, her old job, head of the Air Quality Division, was left vacant. Larson took over the agency in early July.

EPA's Objections

On July 2, 1996, one day after the official 30-day public comment period ended, EPA submitted comments to MPCA indicating that it had problems with the permit. Nevertheless, EPA was willing to negotiate. The letter communicating EPA's concerns to MPCA, signed by Valdus Adamkus, regional administrator in Chicago, stated that EPA believed "the project could be a valuable experiment," that all parties could "work together to resolve any remaining issues," and that discussions should begin immediately.[9]

EPA gave many reasons for not supporting the proposed agreement, but foremost among them was the absence in the proposal of a guaranteed level of direct environmental benefits. EPA was not comfortable with the amount of operating flexibility granted 3M by the proposed permit and FPA. Because SEP was a "condition of both the Minnesota XL and EPA Project XL programs," the agency asked that the permit be modified to ensure that 3M achieved plantwide emissions for *each pollutant* class—VOCs, HAPs, and criteria pollutants—that were no more than 90% of what emissions would have been under otherwise applicable requirements. In determining the latter standard, 3M must assume that all the existing controls on the facility's unregulated units would have been operating; the company would receive no credit for its past voluntary emission reductions. Failure to achieve this level of SEP for *any pollutant* class would be grounds for termination of the XL permit and a "hasty return" to the otherwise applicable requirements.

Overall, regarding SEP and related factors, EPA was seeking several things:

1. that the permit would last only five years, thereby requiring a renewal;
2. an annual, source-by-source regulatory analysis to verify that the entire facility was not emitting more pollutants than allowed by applicable regulations (the proposed permit had asked for only two such reviews during the first five years);
3. a complex source-by-source daily measurement program to enable 3M to show that it was attaining SEP at the plant;
4. SEP for every criteria pollutant in addition to the VOCs and HAPs, assuming that 3M would maintain the voluntary controls that were now in place; and
5. failure to achieve SEP for any one pollutant would be grounds for immediate revocation of the XL permit by EPA, with 3M having no right to appeal or negotiate the decision.

The proposed permit and FPA required three reviews during their 10-year life to ensure that 3M was achieving the environmental performance goals established in the FPA and to determine if changes in the permit were needed in light of the then otherwise applicable requirements. EPA asked instead for annual regulatory reviews to verify compliance by the Hutchinson facility with the agency's definition of SEP for every pollutant and, if need be, to adjust the caps as a result of changes in environmental regulations or changes in the regulatory status of any of the production units at the facility (e.g., modifications in a grandfathered line that triggered regulatory requirements).

EPA was particularly concerned that 3M, while operating under caps that were well above current actual emission levels, would be able to make modifications at the plants that would escape the usual regulatory constraints. This concern was exacerbated by the fact that major changes were taking place at the Hutchinson facility, and little information was available about future operations. New products and production lines were being introduced, and old lines were being closed, making it virtually impossible to predict what regulations might apply to yet to be determined new equipment. The 10-year XL permit originally proposed would have allowed 3M the flexibility to make process changes during the life of the permit without being subject to permit review. This provision was especially bothersome to EPA's air quality program.

EPA also wanted stronger enforcement provisions in the proposed XL permit: "The permit/FPA should be clarified to make plain the consequences of failure to achieve superior environmental performance."[10] The agency argued that "violation ... should result in unilateral termination of the permit/FPA by MPCA and/or EPA and a speedy return to compliance with the

Clean Air Act." In addition, EPA cited several provisions of Title 5 of the Clean Air Act that it could not waive. These included limiting the term of the permit to five years and continuing the "line-by-line record keeping and monitoring, and reporting as required by otherwise applicable requirements." EPA also stated that it would not waive the monitoring, recording, and reporting requirements of the Resource Conservation and Recovery Act and of the Emergency Planning and Community Right-to-Know Act. Monitoring data had to be recorded daily and be included in quarterly reports. The original proposal had envisioned using 3M's environmental management system (EMS) to assure compliance and thereby eliminate some of these reporting requirements. EPA also demanded that "any element of the EMS designed to assure compliance" be approved not only by MPCA but by EPA as well.

Opposition from the Natural Resources Defense Council

Opposition to the agreement also came from national environmental organizations, in particular from the Natural Resources Defense Council (NRDC), which considered the allowable emissions cap inconsistent with the spirit and the letter of the Clean Air Act.[11] The MPCA XL team thought that the local environmentalists on PPC had been in communication with NRDC about what was taking place. In fact, the press of other activities had prevented them from keeping NRDC fully apprised of negotiations on the proposal. The first inkling MPCA had that NRDC was concerned occurred during a presentation about Project XL that Williams made to the President's Council on Sustainable Development in Washington, D.C., in early June 1996. NRDC's executive director asked many questions about the proposed agreement, and NRDC later submitted formal comments during the public comment period.

In its written comments, NRDC said that the VOC emissions cap "would allow the facility to emit more tons of VOCs than all but the 70 most polluting facilities in the U.S."[12] NRDC maintained that the proposed agreement seemed to offer only theoretical reductions instead of requiring actual environmental improvement. The North Plant was already achieving control efficiencies of 97.6–99.6%, which made the best available control technology (BACT) analysis done by MPCA, based on 95% emissions reductions, incorrect. If the higher standard (more than 97.6% reduction) was applied, allowable emissions at the North Plant under BACT would be 1,033 tons of VOCs, not 1,810 tons. However, NRDC argued, because the North Plant was being phased out, its emissions should not count *at all* in calculating the emissions caps. The only emissions that should count were those of the South Plant. Thus, the caps should be set at no more than 1,468 tpy of VOCs.

In its comments, NRDC also maintained that pollution prevention analysis at the plant should not be left to 3M.[13] Rather, it should be done by stakeholders and outside technical consultants, and the results should be used in setting the goals for VOCs and HAPs.[14]

NRDC did not actually threaten litigation, but EPA and 3M were concerned about this possibility, and this concern made them less willing to proceed. Neither wanted the unpleasant publicity or expense.

MPCA's Reaction

The MPCA XL team thought that NRDC's comments were misdirected and off target, primarily because they did not fully understand the proposal. MPCA believed that EPA's comments made it plain that the agency did not support the innovative, experimental nature of the permit.[15] Consistency with the Clean Air Act was more important to EPA than allowing such experiments; XL had to fit in with EPA's traditional air quality compliance program. In MPCA's opinion, EPA's position eliminated most, if not all, of the flexibility and innovation that XL was supposed to encourage.

MPCA had accepted the fact that innovations often took several years to work out and had therefore agreed to a longer time frame for measuring the overall environmental benefits of the project. In MPCA's view, the EPA position was based on concern about the precedent-setting potential of the proposed agreement with 3M.[16] EPA seemed to be using the 3M–Hutchinson pilot to set national policy, sending a message about how far it would go with regulatory reform initiatives.

MPCA's reaction to EPA's comments also was prompted in part from disappointment that the 3M–Hutchinson XL project was not close to agreement and that EPA was not granting the state agency as much decisionmaking authority as it thought it had when it was selected as a Project XL participant. EPA's comments on the Minnesota group's proposal appeared to be merely an opening gambit in the agency's negotiations with 3M and MPCA. MPCA felt that it was rather late in the game for an opening gambit.

3M's Reaction

3M gave its perspective in a memo written on July 25, 1996 (3M Company 1996b). The memo defined SEP as "simply performance (measured in a number of different ways), which is better than required under existing regulation." 3M pointed out that the Hutchinson facility had already demonstrated SEP under the existing regulatory structure and expressed the belief that the proposed XL permit would "facilitate the development of new ways to continue and perhaps surpass the 'superior environmental performance' of the past using common sense alternatives to the existing laws and regulations."

The memo cited several elements in the proposal that ensured environmental performance at Hutchinson that would go well beyond compliance:

1. the VOC cap was 86% below the current allowable emissions level and 15% below what would be allowed if New Source Performance Standards were applied to all 12 tape coaters instead of just three, as was then the case;
2. the cap on HAPs limited these emissions when no such limits existed under current regulations and years before such regulations (maximum achievable control technology, or MACT, standards) were expected, especially for adhesive tape production;
3. the experimental risk assessments of chronic and acute HAP exposures and the operating restrictions imposed by the XL permit to avoid acute HAP exposures were applied to Hutchinson years before any standards would be in force under the Clean Air Act; and
4. the periodic regulatory reviews required by the permit would assess the ongoing environmental performance of the facility compared with the then-applicable command-and-control regulations and, if necessary, lower the caps to ensure SEP. Most of the VOC and HAP emissions sources at the Hutchinson facility were grandfathered under the Clean Air Act and were therefore not regulated at all. The plant's voluntary controls had already gone an enormous distance beyond compliance.

The 3M memo complained that EPA's July 2, 1996, objections to the proposed XL permit were not giving the company "credit for this exemplary performance." The memo said, "We should not be penalized through the regulatory assessment process for implementing this aggressive and forward looking program." The memo noted that 3M was "now accelerating a transition from the era of compliance toward an era of sustainable development." Cited as an example of this move were the corporate goals for 2000 to reduce all releases to air, water, and land to 10% of 1990 levels, and the generation of waste to half the 1990 total. Achieving the goal for air pollutant emissions at Hutchinson, for example, would take VOC emissions levels to below 1,500 tpy, relative to the 1995 level of 2,307 tpy. This goal would be quite difficult to achieve, but trying to achieve it would drive the facility to SEP independent of any credit for past voluntary emissions reductions.

Last, the memo cited the fact that embedded as enforceable parts of the proposed XL permit were the company's Air Emissions Reduction Program requirements that best available control technology be installed on all VOC sources emitting more than 100 tpy and that controls be used on all new VOC sources emitting more than 40 tpy. The details for implementing these permit requirements would be developed in the Hutchinson facility's new advanced EMS. The 3M memo also revealed some of the near-term plans for changes at the Hutchinson facility as magnetic audiovisual tape production was phased out.

This memo was meant to be a signal to EPA that time was running out for 3M. The company had to make decisions about the audiovisual plant and needed a speedy resolution to the negotiations.

The failure of the Minnesota contingent to broaden an initial group of supporters and gain wider backing—especially from influential parties at EPA and the national environmental community, with the ability, in effect, to veto the agreement—was a serious roadblock to reaching an agreement. How serious this obstacle was and whether it could be overcome will be explored in the next chapter.

Notes

[1] Based on 1996–1997 interviews with MPCA staff. See the Acknowledgements for a description of these interviews.

[2] Letter from T. Zosel, manager of environmental initiatives at 3M, to S. Herman, assistant administrator in EPA's Office of Enforcement and Compliance Assurance, expressing concern about reliance on "standard operating procedures" rather than "innovative thinking," Oct. 16, 1995.

[3] Based on 1996 interviews with regional EPA staff. See the Acknowledgements for a description of these interviews.

[4] Notes by MPCA staff on the meeting with staff from EPA Region 5 and EPA headquarters in Chicago on June 15, 1996.

[5] Actually, the 1991–1995 five-year average is below the cap.

[6] The 34,000-ton emissions limit was at full utilization of all production lines and assuming controls on only three regulated lines.

[7] 1996–1997 interviews with MPCA staff.

[8] Ibid.

[9] "Minnesota XL Permit/Final Project Agreement Comments," letter from V. Adamkus, EPA Region 5 administrator, to C. Williams, MPCA commissioner, July 2, 1996.

[10] Ibid.

[11] NRDC, "Comments on Minnesota XL Permit No. 96-01," June 29, 1996.

[12] Ibid.

[13] Ibid.

[14] Also see letter from C. van Loben Sels, senior project analyst at NRDC, to the Keystone Center, March 21, 1996; letter from D. Hawkins, senior NRDC attorney and van Loben Sels to F. Hansen, deputy administrator of EPA, and D. Gardiner, assistant administrator of EPA's OPPE, on poor initial results on superior environmental performance, July 1, 1996; NRDC, "Comments on the Initial XL Projects (Intel, Merck, 3M–Hutchinson)," July 2, 1996; and letter from van Loben Sels and E. Dorsey to F. Hansen with attachment, "The Model Plan for Public Participation," July 18, 1996.

[15] 1996–1997 interviews with MPCA staff.

[16] Ibid.

7

Trying To Overcome Deadlock: The Practical Impediments

Though there were difficulties in forging a broader coalition that included the U.S. Environmental Protection Agency (EPA) and the national environmental organizations, it did not mean that the supporters of the 3M Company's XL project gave up entirely. Even as 3M lost interest, EPA suggested a new approach, an alternative it called the "comparative action test." The agency worked with the Minnesota Pollution Control Agency (MPCA) to restart the negotiation process. This time around, new actors with a fresh perspective were involved; not bound by earlier thinking, they hoped that they might be able to hammer out an accord. But they still were unable to overcome the deadlock that had developed. In this chapter, we describe their ongoing efforts to overcome the impasse. We ask why the negotiations ultimately broke down. Though a variety of perspectives can be applied, two impediments are emphasized: the proposed terms of the agreement (the substance) and the decisionmaking process. Before proceeding to the Intel, Weyerhaeuser, and Merck comparisons, we conclude this chapter with a summary of our analysis of why the 3M negotiations ended in deadlock.

Continued Hope for a Deal

Although willing to give the negotiations with EPA one more try, 3M staff were disillusioned. They could not wait to make changes at the Hutchinson,

Minnesota, facility and might have to proceed without an XL permit in place.[1] With time running out, 3M's XL project team wanted MPCA to appeal to President Bill Clinton. Given the changes that 3M was making at the Hutchinson facility, 3M's staff felt a great urgency.

In one of his last actions as the MPCA commissioner, Charles Williams obtained support for the Hutchinson agreement from Minnesota's congressional delegation, including both senators, Paul Wellstone and Rod Grams.[2] He also received backing from several environmental advocacy groups in Minnesota and from the local Hutchinson community.[3] The commissioner's letter to Clinton said that MPCA had tried to cooperate with EPA. It had met and communicated often with EPA but was surprised by EPA's objections. Williams believed that the 3M pilot fit the criteria that the president had established for XL. Echoing the president's remarks when he inaugurated Project XL, Williams maintained that the pilot set the bar for environmental achievement high and gave 3M flexibility to choose how it would jump over it. The 3M proposal was protective of human health and the environment and reduced the administrative burden on 3M and the state. Williams sent copies of the letter to Minnesota's congressional delegation and its governor, to Vice President Al Gore, and to Clinton's special assistant for environmental affairs.

Though the letter did not elicit a substantive response from these parties, MPCA staff continued to believe that they might be able to arrange a deal. EPA had communicated informally to MPCA that everything in its comments about the permit and the proposed final project agreement (FPA) was negotiable. MPCA staff believed that agreement on many issues was close. If those issues could be resolved, the outlook would improve for reaching agreement on the more difficult issue of the appropriate level of superior environmental performance (SEP).

Another Meeting with EPA

To continue with the negotiations, MPCA arranged another meeting with EPA, this one on July 30, 1996, in Bloomington, Minnesota, a suburb of Minneapolis.[4] At the meeting, David Ullrich, deputy administrator of EPA Region 5, stated that the agency was eager to have a flexible agreement but needed to deal with the legal impediments. From the agency's perspective, there were many environmental assurances and advantages provided for by existing statutes. Before the agency would forgo the way it typically implemented these statutes, there had to be very clear guarantees of SEP.

Ullrich reiterated EPA's position that, as a matter of principle, Project XL pilots should not proceed outside the framework of the Clean Air Act and that because the proposed agreement was in very important respects not compatible with the act, it could not be carried out within its parameters. For example, under the Clean Air Act, process modifications at the site required EPA's preapproval. Trades across the plant's regulated and nonregulated sources of volatile organic compounds (VOCs) ordinarily would not be allowed, nor would trades among different sources of hazardous air pollutants (HAPs). Moreover, Ullrich contended, the test for SEP should be extended to cover such criteria pollutants as sulfur dioxide and particulate matter.

Ullrich reminded those present that the Clean Air Act mandated technological determinations of pollution control adequacy on a source-by-source basis rather than permitting a facilitywide bubble. He maintained that defining SEP to be the achievement of an overall goal for an entire facility was not consistent with this law. The proposed overall cap on VOCs and HAPs at Hutchinson did not conform to the best available control technology (BACT) and maximum achievable control technology (MACT) standards in the Clean Air Act. Ullrich argued that SEP achievement would thus have to be assessed in the only manner permitted under the act—one source at a time.

EPA, according to Ullrich, had other problems with the proposed agreement. The agency wanted not only the permit to be enforceable, but also the elements of the FPA that would be used to determine SEP, including parts of 3M's environmental management system. If 3M did not comply with these aspects of the FPA, Ullrich asserted, it should be grounds for immediate termination. The agreement should be abrogated, and there should be a return to conventional regulation. Ullrich also mentioned that under the Clean Air Act, regardless of what else occurred, an XL permit could be in effect for no more than five years. In addition, he maintained that EPA was concerned that under the caps in the proposed permit, 3M could introduce new equipment with substandard controls without permit violation, given the large amount of voluntary reductions it had made, and that 3M should not be granted credit for its past good behavior.

Ullrich noted that EPA always had to anticipate that there might be lawsuits as a result of its decisions. Therefore, the agency wanted to be sure that it was proceeding on a solid legal foundation. EPA understood that its position would make it harder for 3M to innovate; plant engineers would have to spend time on compliance activities that they otherwise could spend pursuing the development of new technologies. However, whether or not the proposed permit was legally defensible remained the most important question for EPA's legal counsel and its program offices. EPA's purpose was not to make compliance with environmental protection ordinances easier for 3M.

3M's response was that for an experiment to work, EPA had to be more flexible than it was showing itself to be. The administration's intention in setting up Project XL was to bring about a paradigm shift that would reinvent government by granting regulatory flexibility to those who could demonstrate SEP. The purpose was to empower people at the local level to innovate so that both the environment and communities would be better off. The company had always been an environmental leader. It operated its facilities around the world in accord with corporate goals to not just meet minimum standards but go beyond them. It had spent $44 million on voluntary controls at the Hutchinson facility alone. In addition to protecting the environment, 3M's aim was to keep good jobs in the community. 3M felt that it was being penalized for having voluntarily taken environmental initiatives in the past and that this was not fair.

An MPCA Pilot Project Committee (PPC) member made the point that EPA should consider the learning that would take place when a facility that was bound to overall performance goals was given flexibility. As long as a baseline guarantee existed that protected the environment, a facility should be able to experiment with setting its own goals. It should be allowed to experiment with a facilitywide system, such as mass balance analysis, without these measures becoming part of an enforceable agreement with the government. The government would determine after the fact whether the experiment was successful, meaning whether it stimulated better results for the environment and the economy than otherwise would have been the case.[5]

EPA staff, however, said that they were concerned about 3M's potential backsliding. If the permit allowed 3M to emit 3,000 tons per year of HAPs and it currently emitted 1,100 tons, what was to prevent 3M from moving closer to the 3,000-ton limit? As a regulatory agency, EPA could not rely on unenforceable corporate goals to keep 3M at or below its current emissions levels. Caps had to be reassessed annually, and the agency had to have the right to terminate the agreement if 3M exceeded its current emissions. EPA asked whether it was possible that 3M would fail an "actuals-to-actuals" emissions test. Were not the new production techniques it was planning far more efficient than the old? Would not the changes 3M was planning to make—pulling out old lines and putting in new lines—keep it comfortably below regulatory minimums?

3M responded that it was committed to expanding the company at a rate that was 6% greater than the U.S. economic growth rate and that, as it grew, it faced risks in reformulating products and introducing new manufacturing lines. Though it was committed to improvements such as solvent-free formulas, it could not predict whether these improvements would keep pace with its need to grow the company. In the long run, such innovations might very well function better than expected. But in the short run, they might result in temporary emissions increases. 3M said that it appreciated EPA's efforts and that it would be willing to discuss various ways of demonstrating success and enforceability in the future but for now what EPA proposed was unfeasible. Ullrich urged 3M not to walk away from the negotiations and to continue to work on a proposed XL agreement, but 3M refused to give any assurances that it could, given the positions EPA had adopted.[6]

More Discussions

After the July meeting, discussion continued among EPA, MPCA, and 3M.[7] EPA slowly made compromises, backing off from positions it had taken, but 3M continued to have objections.[8] 3M would be forced to accept a level of regulation at the Hutchinson facility that was stricter than current regulations, increasing the chances that its XL permit could be revoked and that the facility would have to return to conventional command-and-control requirements. Hutchinson engineers would be effectively participating in a dual regulatory process, and therefore they would not be able to take full advantage of a performance-based framework. From July 30 until August 16, there was a period of intense activity. MPCA tried to act as mediator and attempted to extract compromises from 3M and EPA. 3M agreed to accept an enforceable commitment in which it would outperform total emissions of VOCs and HAPs otherwise allowed by applicable requirements by 10%.[9] 3M also agreed to language in the permit requiring it to prepare and submit pollution prevention reports to MPCA and 3M stakeholders annually, to hold meetings twice a year with 3M stakeholders, and to add Project XL emissions data to the company's website, thereby making these FPA commitments enforceable.

David Gardiner, assistant administrator of EPA's Office of Policy, Planning, and Evaluation (OPPE), asked 3M to agree on the principle that it would not use "the impressive voluntary steps it has taken in the past to offset below-standard emission controls in future plant expansions."[10] To ensure that this principle was adhered to, EPA wanted an "actuals-to-actuals" test. 3M would have to compare actual annual pollutant emissions with a calculation of what actual emissions would have been under otherwise applicable requirements assuming voluntary controls remained in place. Recognizing that time to negotiate an agreement was running out, Gardiner followed up his letter of August 8 with a conference call the following day, involving staff from EPA, MPCA, and 3M.

David Sonstegard, head of 3M's Environmental Technology and Safety Services Division, expressed the hope that after the exchange of views it would be possible to "finally resolve the two substantive policy issues which remain unresolved and MPCA can issue the 3M Hutchinson XL Permit."[11] These two issues were the definition of SEP and the legal mechanism to approve Project XL. Sonstegard reiterated 3M's view that EPA's position with regard to SEP was not in the spirit of what the company "perceived to be the purpose of Project XL—that is to let companies that demonstrated histories of excellence and leadership propose pilots to experiment with the current environmental regulatory system with hopes of creating a path to regulatory reform." Nevertheless, 3M still believed that XL was "an important program" that could "be a meaningful path to a true performance-based environmental regulatory system." Arguing that "small steps toward reform are better than no steps," Sonstegard offered to "incorporate EPA's definition of SEP into the mechanics of the permit as requested by Gardiner."

To this end, Sonstegard proposed an amendment to the XL permit that compared for a given year emissions of VOCs and HAPs from all new sources that began operation after September 1, 1996, with allowed emissions under federal and state regulations. Sonstegard noted that the 3M–Hutchinson permit was far from what 3M had envisioned and proposed as a pilot. The permit now contained EPA's definition of SEP, and it had workload and paperwork requirements that equaled or exceeded those required under the current system. With these final changes, however, Sonstegard said that 3M considered "all outstanding issues that have been raised by the EPA as addressed." However, he added that EPA's latest proposal to issue 3M's XL permit under the Clean Air Act Title 5 permitting program was a "last-minute change" from previous EPA positions to employ site-specific rulemaking. Not obtaining a site-specific rule would be a deal-breaker for 3M. Sonstegard ended by saying, "I now look forward to a final decision and written confirmation by EPA on this matter by Friday, August 16, 1996."[12]

In a memo to the MPCA on August 16, Jon Kessler, director of the Emerging Sectors and Strategy Division in EPA's OPPE, sketched EPA's ideas for conducting a unit-by-unit analysis of pollutant emissions to test for SEP at Hutchinson. Kessler's memo introduced an approach whereby the Hutchinson facility could obtain SEP credits for voluntary improvements on all its grandfathered emissions units combined (i.e., unregulated sources) made after September 1, 1996. These credits, if sufficiently large, could compensate for underperformance at some regulated units and yield an overall positive assessment that SEP had taken place. The company also could obtain credits for overcontrolling any of its regulated emissions units.

More discussion was necessary because these proposals for incorporating explicit SEP guarantees into the XL permit had grown increasingly complex. The MPCA, however, started to lose patience. On August 27, the agency's new commissioner, Peder Larson, sent a letter to EPA Administrator Carol Browner announcing MPCA's intention to suspend work on the 3M XL project. Larson stated that EPA's August 16 proposal (Kessler's memo) contained "a restrictive and burdensome permit condition for guaranteeing superior environmental performance up-front." He noted that on August 20, both 3M corporate staff and Hutchinson plant staff *rejected* EPA's proposal—a decision supported by MPCA. Larson pointed out that "we never envisioned requiring prescriptive permit conditions which render the experimental nature of XL moot." Larson said that MPCA was considering suspending its participation in XL unless significant differences between its expectations of and experiences with the project could be resolved.[13]

In an August 29 letter, Gardiner replied for EPA's OPPE. He reviewed what he believed to be agreements that already had been reached: a streamlined regulatory review process to facilitate more rapid changes at the Hutchinson plant; flexibility for 3M to achieve facilitywide air pollution standards in the best ways it saw fit; and flexibility to verify compliance using 3M's environmental management system, thereby reducing record keeping and reporting requirements. Gardiner said that EPA would agree to 3M's request for a site-specific rule, but only as long as the agreement would "provide the people of Minnesota with the same level of environmental and public health protection that they are now guaranteed by the Clean Air Act and other environmental laws." Consequently, Gardiner felt that "additional implementation language" was needed to ensure that this "vital goal" was met.

Gardiner's remarks at first glance seem quite puzzling, because 3M's Hutchinson facility was already operating at a level of environmental performance that was far superior to what was mandated by the Clean Air Act as a result of extensive investments to prevent or control emissions by the many unregulated sources at the plant. However, new emissions units and sufficiently modified grandfathered sources would come under Clean Air Act regulations, and EPA was concerned that such sources would somehow be less strictly controlled if Hutchinson operated under its proposed caps and without the SEP requirements EPA sought. Gardiner reiterated his support for EPA's proposed language to provide this guarantee for "the entire facility, rather than simply to new units." Furthermore, he noted, EPA's language also confirmed "that the facility will meet the protection requirements during the term of this XL project, rather than by trading future performance against voluntary controls taken prior to the start of Project XL."

Sonstegard of 3M responded on September 5.[14] He dismissed Gardiner's claims about what had been accomplished. He maintained that EPA was not acting in a timely enough fashion to satisfy 3M's business needs. He wrote that "business considerations now lead us to conclude that we must move forward immediately with efforts to obtain traditional permits and to shelve the Project XL experiment at 3M Hutchinson." The decision, he said, was based on "operational and timing considerations" because new products and processes had to be introduced at the facility to meet 3M's commitments to its customers and employees. Although Sonstegard seemed to eliminate the possibility of further negotiation, he did invite the recipients of the letter to meet with him and the Hutchinson plant manager to explain in greater detail how the company arrived at its decision. He expressed his belief that "3M's decision rationale, as based on our investment, experience and business needs, would greatly benefit the evolving development of the XL program."[15]

Final Efforts To Resuscitate the Project

Sonstegard's letter ended the formal Project XL negotiations among the parties. Nonetheless, MPCA and PPC continued to try to resuscitate the process.[16] Their first action was to send letters to Browner reviewing what they regarded to be the environmental benefits of the XL permit and FPA originally proposed in early June and urging EPA to reconsider its position. They worked with the Minnesota Attorney General's Office, which sent a similar letter to Browner, and with six Minnesota members of Congress and one of the state's U.S. senators, who also sent a joint letter, in this instance to Clinton, in support of the 3M XL project. Neither Browner nor Clinton, however, gave a meaningful response.

Nonetheless, MPCA, with backing from PPC, continued to promote high-level discussions. In response to Sonstegard's invitation, Gardiner and Lisa Lund from EPA OPPE visited the Hutchinson site on September 18, 1996.[17] The EPA officials toured the North and South plants and met with 3M managers, who gave them a history of 3M's efforts to control emissions at the facility and explained why EPA's proposed emissions unit-by-unit test for SEP would not work. The result of this meeting was that the EPA officials thought they could improve the test, and they said that they intended to further develop it in an attempt to address the issues and concerns 3M raised. 3M indicated that it was willing to continue with the discussions but only if EPA created a team consisting of high-level people from EPA and the MPCA commissioner and only if these discussions were off the record and did not involve stakeholders and the public.

EPA, 3M, and MPCA agreed to have additional discussions under these conditions. These confidential discussions continued almost to the end of December 1996. The discussions were confidential, in part, because of 3M's reluctance to raise any false hopes because it was skeptical that an agreement could be reached.

Gardiner, Sonstegard, and Larson became the principals in these continuing negotiations and the key decisionmakers on a new team that was trying to take a fresh look at the issues. Along with key staff, they planned to meet again in Washington, D.C., during October 1996, to review progress that had been made and focus on concerns about EPA's proposed test for SEP and other outstanding issues. Between September 19 and the October 10 meeting in Washington, EPA, MPCA, and 3M were able to reach conceptual agreement on some issues.[18] The most notable were those concerned with implementing the new MACT regulations for magnetic tape production at the North Plant.[19] EPA agreed to "bubble" the MACT requirement, thereby relieving 3M from the need to apply MACT standards to the facility's smaller, uncontrolled sources, such as the plant's compounding area. Despite this progress with the negotiations, the parties still did not have a resolution of the major problems relating to an emissions test for SEP.

EPA's Test for SEP

The test that EPA devised was based on the following premise. SEP meant that a facility had to do better than it otherwise would have done without XL. In the 3M–Hutchinson case, the agency assumed that absent an XL permit, the facility would have continued to operate its large number of voluntary emissions control systems.[20] Thus, SEP had to be measured from an actual emissions baseline, but one adjusted for changes in production levels after September 1, 1996. Any pollution prevention or control initiatives taken after that date, whether on regulated units or grandfathered units, would be credited toward SEP—just as any production-level-adjusted emissions increases, for whatever reason, would be debited against SEP.

EPA did not define the caps on the basis of units of production. Given the complexity and variability of 3M's product mix, this would have been difficult to accomplish. Consequently, to accommodate increased production, the caps had to be set higher than actual emissions levels. That they were set this way opened the door for EPA's XL participants to imagine worst-case scenarios. Gardiner expressed these concerns in a September 27 letter to Sonstegard. He stated that EPA was seeking assurance that performance under the agreement "would be judged relative to a baseline of 'actual' rather than maximum 'allowable' emissions." He went on to say that the focus on emissions that would have occurred without XL should be on the amount of pollution *actually released*, not a hypothetical maximum amount that *could have been allowed*. Otherwise, projects portrayed as superior might allow companies to emit more than if they continued to comply with existing requirements.[21]

In addition to this disagreement over the baseline from which to measure SEP, there were differences over several other aspects of EPA's proposed amendment of the permit. EPA wanted the SEP test to be applied not only to VOCs and HAPs, the real focus of the 3M project, but to the criteria pollutants as well. The federal agency also wanted the tests to be performed annually and to have failure to pass the test for any of the pollutants be grounds for termination of the XL permit, although such termination would not be mandatory or even likely. 3M and MPCA wanted less frequent tests. Aside from the added compliance burden imposed by annual testing, both believed that less frequent testing would give 3M– Hutchinson engineers the time needed to try innovative approaches to environmental and public health protection. Annual tests would discourage risk taking. Hutchinson was an attainment area, and the plant's criteria pollutant emissions, though not negligible, were not a major concern to MPCA. VOCs and HAPs—essentially all a subset of VOCs—were a concern and, in the views of 3M, MPCA, and the stakeholders, were where SEP should be primarily judged.

The Comparable Actions Test

Around the time of the October 10 meeting in Washington, EPA started to refer to its evolving test for SEP as the "comparable actions test" (CAT). At the October 10 meeting, 3M and MPCA made it clear that they did not believe that EPA's approach was consistent with the principles of Project XL.[22] Another meeting was planned for the end of October, at EPA's Office of Air Quality Planning and Standards (OAQPS) at Research Triangle Park in North Carolina. Unlike the meeting in Washington, where no technical support staff from any of the three organizations was present, the meeting in North Carolina on October 28–29 involved several engineering and legal experts representing the participants.[23] The meeting also involved key staff from MPCA, including the air permit engineer working on the 3M project. Because the purpose of this meeting was to focus on the details of the permit and FPA as well as the CAT, Gardiner, Sonstegard, and Larson were not there.

After the October 28–29 meeting, EPA staff worked with MPCA and 3M to try to improve the CAT. Numerous conference calls took place, in which about 30 people from MPCA, 3M, and EPA participated. Lydia Wegman, assistant administrator of OAQPS, facilitated the calls. Gail Lacy, an environmental engineer with OAQPS in North Carolina and Peggy Bartz, an air permit engineer at MPCA, did most of the technical work. Bartz would test various versions, and when the results did not seem to make sense, she would go back to the EPA technical staff, who would refine the language so that the results were closer to what the parties wanted. The subsequent drafts eliminated some of the inconsistencies; in Bartz's opinion, however, some ambiguities remained. In the process, the CAT became more complicated.

Nonetheless, MPCA and 3M agreed to let EPA continue to try to refine the CAT.[24] EPA produced a final written version of the CAT in December, and MPCA received a copy of it on December 11 (Ronchak 1996). The CAT applied only to adhesive tape production. Because 3M was planning to phase out audiovisual magnetic tape production, EPA had agreed to place a bubble over that operation at the North Plant.[25]

In its attempt to cover a wide range of unpredictable circumstances at the Hutchinson facility, the December 1996 version of the CAT was complicated and difficult for 3M and MPCA to fully comprehend (see Appendix A for the current CAT). Had 3M been interested in pursuing this approach, the CAT would have needed further refinement. The basic approach was little changed from ideas Kessler had sketched out in his August 16 memo, but the details, as 3M and MPCA saw it, were hard to pin down.[26] What 3M and MPCA did not like was that the CAT provided a performance standard that was variable and likely to change from year to year. The CAT required both 3M and MPCA to undertake complex and costly source-by-source calculations that both organizations thought were unnecessary.[27] 3M was opposed to annual compliance reviews because some of the innovations it was considering might take time to work out. The company's environmental managers estimated that EPA's proposed compliance requirements under the XL permit would be more costly than under the existing regulatory system.[28]

On December 9 and 10, 3M managers, in accord with their agreement with EPA and MPCA, presented to the Hutchinson plant managers and engineers the latest version of the FPA and the XL permit, which now included the CAT (Ronchak 1996). The company would not move forward with the project unless the site managers and engineers found the new version workable. Gardiner, Sonstegard, and Larson held a conference call on December 12. Sonstegard told EPA that although the plant managers and engineers had not arrived at any definite conclusions, they were cool to the approaches being taken. 3M doubted if its plant managers and engineers would be able to work effectively with this new version by the agency.

On December 18 and 20, key EPA, 3M, and MPCA staff held additional conference calls. EPA outlined three options (Ronchak 1996). The first was to accept, perhaps after some final fine-tuning, the CAT, complex though it might be, and then implement it. The second was to create an alternative to the CAT that was a less exact but a simpler approximation of what the facility would actually achieve under the Clean Air Act without Project XL. The third option was to abandon the Hutchinson project entirely. EPA asked 3M to choose one of these three options. On December 24, 3M notified MPCA and EPA that the Hutchinson management had decided to abandon XL.

As a private corporation involved in this process, 3M initially risked a great deal. 3M had invested about $1 million (including legal fees of more than $300,000) in Project XL.[29] It became involved when few other companies were willing to do so, when it was unclear what the results would be and whether there would be any benefits. In late August, when it became increasingly clear that the gains that 3M was seeking would not be forthcoming in a timely manner, the company felt that it had to withdraw. Nonetheless, it was unwilling to completely sever its ties to XL, perhaps even hoping that something positive could be salvaged, or possibly it simply wanted to maintain a good relationship with EPA. Consequently, 3M continued to try to negotiate through December, but in the end this attempt failed.

3M may have gained some benefits from participating in Project XL by solidifying its ties to MPCA and demonstrating the advantages of a team approach in its future dealings with the state agency. The company also obtained an independent look at its operating procedures from MPCA's professional staff, including MPCA's chemical engineers. The greater familiarity that MPCA's permitting engineers had with the facility made it easier for 3M to obtain conventional permits on its new lines at Hutchinson. 3M also did not make concessions to EPA that it could not abide by, simply for the sake of reaching an agreement. It showed that there were limits to how far it would be willing to go to obtain an XL permit. (See Appendix B for a description of the next steps 3M took at Hutchinson.)

EPA started the negotiations with a tougher stance than where it ended. It started to back off when it saw that 3M was not likely to continue to negotiate if it adhered to these positions. Trying to overcome deadlock and starting again with new actors who did not have previous experience with the earlier agreement did not work. The obstacles to an agreement were too many and were not overcome.

Substance and Process: The Practical Impediments

Why did these negotiations break down? A variety of perspectives can be applied—transaction costs; the meaning of laws such as the Clean Air Act; the way issues were framed; how issues were selected, discussed, and debated; and the decisionmaking and policy processes on which the parties relied. In what follows, we emphasize the proposed terms of the agreement and the process of how decisions were made.

Substance

There were a number of reasons the participants failed to reach an agreement, some unique to the 3M–EPA negotiations and others connected to problems in Project XL itself. The premise of XL was that in exchange for SEP, regulated entities would be granted greater flexibility (Chapter 2). Project XL–Minnesota was plagued by the question of how to apply this concept to the 3M–Hutchinson facility and how to guarantee it over the life of a long-term permit. The answers the participants gave differed. EPA argued that if facilities are granted flexibility, they must provide a *guaranteed* level of SEP—the greater the economic benefit, the greater the SEP required. 3M disagreed with EPA about how much SEP should be required and about the way in which the company would have to demonstrate in advance that it would achieve SEP.

Thus, one of the main reasons the parties failed to reach an agreement was because, in EPA's view, 3M did not offer enough guaranteed SEP to justify the flexibility the company was seeking. Although EPA was looking for guaranteed direct benefits, MPCA and PPC were looking for systemic benefits that could be realized over a longer period of time (Chapter 3). They took the position that the XL project was an experiment whose purpose was to test new ideas and to leave open the possibility of failure, and thus a guarantee that SEP would result was not absolutely necessary (Dorf and Sabel 1998).

EPA's goal for Project XL was to find ways to tweak the system for incremental improvement. Environmental gains in the here and now were paramount. The quid pro quo was a framework to be applied in particular situations, not a broad principle to be used for the purposes of changing regulation as it had been carried out to this point. MPCA, PPC, and 3M, conversely, were eager for broader institutional change that could move the United States toward a new approach to regulation and to environmental management.

Though the Minnesota participants created an innovative permit and FPA, which encouraged pollution prevention, they were not able to anticipate and overcome the substantive barriers they faced (MPCA 1998). In particular, determining what allowable emissions were and what they would be in a facility that was undergoing substantial changes and modifications in response to market opportunities and technological developments was a barrier to an agreement (Chapter 4). MPCA and PPC considered SEP to be a many-faceted concept that included pollution prevention, advanced environmental management systems, multistakeholder involvement, simpler permits, greater transparency, public understanding, and cost savings. It involved more than a few emission numbers. It required that pollution control agencies look holistically at the facility and encouraged pollution prevention approaches (Chapter 5).[30]

Minnesota was willing to take more risks than EPA was (Chapter 6). MPCA, PPC, and 3M regarded EPA's approach of requiring in the XL permit more guarantees of direct environmental benefits as diminishing the learning that could take place if Hutchinson were allowed to operate under the more flexible permit designed by the Minnesota stakeholders. The basis for EPA's concerns was unease among its staff and within nongovernmental organizations such as the Natural Resources Defense Council about granting companies tangible economic benefit from regulatory flexibility. EPA staff and nongovernmental organizations wanted to be sure that government obtained tangible environmental benefits in return.

Looking at the 3M–Hutchinson XL proposal, EPA's XL team was troubled by the fact that the VOC and HAP caps allowed 3M to emit more air pollutants than the Hutchinson facility was currently emitting. The EPA team was troubled despite the fact that the facility, located in an attainment area, was able to increase emissions under current regulations, either by increasing production or by diminished control of the plant's many unregulated lines; it was the *appearance* of granting 3M a right to pollute that disturbed

EPA staff and environmental advocates. Many EPA staff members were also concerned that 3M would use the large gap between current actual emissions and the caps to introduce new sources with substandard controls.[31]

The large number of voluntary controls and other emissions reduction initiatives that 3M had applied to its unregulated production lines added to the difficulties in reaching an agreement, difficulties that came to the forefront when EPA offered the CAT proposal. There also were the compliance requirements in the CAT, complicated by the need to monitor an experiment that imposed a costly burden on 3M and MPCA. Finally, the substantial risk to 3M that the XL permit could be terminated contributed significantly to the company's reluctance to continue with the pilot project. Investments made under XL to reduce pollution in the most cost-effective way could leave 3M in violation of conventional regulations if 3M's XL permit were to be revoked.

Process

In addition to these differences about the substance of an agreement, the participants encountered problems related to the process of coming to an acceptable settlement. To begin with, the project suffered from the lack of clear delegation of decisionmaking authority. The customary way in which the participants related, moreover, was linear or sequential, and discussion of problems was fragmented and separated by time, place, and category. The participants did not see themselves as a team working together with a common purpose.

In retrospect, a key mistake was most likely the failure of MPCA and EPA Region 5 to arrive at a memorandum of understanding (MOU) (U.S. EPA 1995a; MPCA 1995c, 1996d) or any type of written agreement that clarified expectations and goals (Weber 1998). An MOU, worked out early in the process, might have prevented some of the difficulties that later developed.

Because EPA had selected MPCA as one of the original XL participants, MPCA thought it had the authority to develop XL permits for Minnesota facilities. The state agency had stated clearly in its XL application that it intended to relieve EPA of administrative burdens by managing XL projects (MPCA 1995b). EPA project managers tried to correct this misapprehension, but MPCA and EPA failed to resolve the issue. The MPCA staff believed that if the only authority EPA gave the state was to help select XL applicants, then MPCA had no more authority than state pollution control agencies that were not XL participants.

The failure to complete an MOU is just one example of the fact that the parties never reached full accord on the details of who was to do what. EPA Region 5 had extensive experience working with MPCA and the 3M–Hutchinson facility, and thus XL participants from Region 5 were best suited to work on the project. Unfortunately, Region 5 did not have the authority to act on its own. Many different EPA staff members in Washington, D.C., and in Research Triangle Park, North Carolina—from OPPE, the Office of General Counsel, the Office of Enforcement and Compliance Assurance, OAQPS, the various other program divisions, and even top management—ended up playing key roles. EPA's view was that Project XL involved national policy issues that could not be properly addressed by a regional office. Nevertheless, Region 5 staff could have played an important role as a stakeholder participating in the design of the project and by acting as a champion for the project within EPA.

At the same time, during 1996, several other similar projects did work closely with their EPA regional offices and with their region's help were able to gain approval of their FPAs (see the next two chapters). However, believing that MPCA had greater decisionmaking authority from EPA than it actually had, neither MPCA nor 3M encouraged Region 5 participation. This situation was not helped by the fact that Region 5 had the reputation for being one of the least flexible of the regional offices.

Within EPA, there was no clear delegation of decisionmaking authority, whether vertically between headquarters and the regional office, or horizontally among EPA's different assistant administrator offices (policy, air, legal, enforcement, and so on). EPA, whose decisionmaking process was one of consensus among its different offices, created a situation where, in essence, any group with reservations about the agreement could veto it or slow it down.

An accomplishment of Project XL–Minnesota was that the local participants did develop common understandings. Though the relationship was difficult to bring about, one did grow among the main participants in Minnesota: 3M, MPCA, and PPC. MPCA acted effectively as an intermediary between its stakeholder group and 3M, and for a time the Minnesota participants functioned as a relatively cohesive unit. Despite some tension, they managed to work together in a manner that the management literature might call boundaryless and virtual, a result that is not new or unusual for other successful collaborative approaches (Weber 1998).

However, during the period when this Minnesota bloc temporarily coalesced, it had little contact with EPA's offices in Chicago or Washington. During the critical months of 1996, when the proposed permit and FPA for the Hutchinson facility were drafted, the Minnesota contingent essentially acted on its own. EPA did not insist on its right to participate, and the Minnesota contingent did not demand that EPA offer its advice and assistance. EPA only became engaged again at the meeting in Chicago (Chapter 6) toward the end of May when MPCA and 3M unveiled the agreement they had reached. By then, many technical decisions had been made. They were based on so many implicit assumptions that it was hard for EPA to understand what had happened.

The numerous assumptions that shaped the agreement were reasonable. There was no single, best solution to the problems the parties faced at Hutchinson, and thus without these assumptions, progress could not have been made. Out of context and from a distance, however, the assumptions were hard to comprehend. They were based on complex trade-offs and tacit understandings that EPA could not have fully appreciated unless it had participated more fully earlier in the process.

The drafters of the permit and FPA were engaging in an experimental effort that they believed was part of the first steps toward forming a new approach to environmental management. They were working in a trial-and-error fashion, in the manner of an experimenter. EPA's expertise and judgement could have been enlisted to support this process, had the agency been willing and able to do so. If EPA's concerns had been heard earlier, paths would not have been taken that appeared irreversible, and the Minnesota participants might have been able to alter their positions to accommodate the federal agency's point of view. Expectations about what could be accomplished under Project XL would not have grown so high among the Minnesota participants, and there would have been less disappointment if the parties had been unable to accommodate EPA's point of view. Some of the transaction costs that were incurred could have been avoided. Indeed, it is possible that EPA's participation as a team member might have led to an earlier termination of the pilot. The project may have been doomed from the start, and if there had been better communication, the participants would have known sooner.

Clearly, the Minnesota group did not put enough effort into bringing EPA into the design of the permit. Telephone conferences proved to be a poor substitute for face-to-face meetings. Furthermore, because MPCA felt that it had been delegated authority by EPA and because 3M preferred to deal with staff at EPA headquarters, the flow of data and other information between the Minnesota participants and Region 5 was impeded. Communication was so poor at one point that Region 5's Air Quality Division temporarily withdrew from the project.

Once the proposal had been presented to EPA, discussion of the issues moved back and forth from group to group in serial fashion, rather than all the participants simultaneously giving their concentrated attention to common tasks and problems. This movement was reminiscent of a multiplayer Ping-Pong game, with misguided shots instead of gentle lobs and a succession of repeated, well-orchestrated engagements. The switching of attention from group to group meant delays and misunderstandings. The proposal moved back and forth not only among 3M, MPCA, PPC, and Region 5 but also back and forth among the many offices in EPA, involving policy and planning, legal affairs, enforcement, air quality, various other program functions, and even EPA's top managers. Compounding the difficulty was the personnel turnover within groups. The succession of new faces and the need to introduce them to the basic issues put a burden on all the parties.

A further complication was caused by the lack of full and complete communication and coordination among environmentalists at the national and local levels and, to an extent, among the different environmental groups with an interest in the outcome. Environmental groups did not have ongoing information about project developments, so when they finally saw the proposed permit and FPA it was far into the process. The permit and FPA were crafted with little opportunity for input from these other players. The many assumptions and trade-offs that went into drafting the permit and FPA were not fully articulated and could not have been easily understood from a simple perusal of these documents. Because of this problem with information flow, environmental organizations were late to receive the information that they needed to make decisions. Having said that, however, it is not clear that bringing national environmental organizations such as the Natural Resources Defense Council more completely into the process would have helped to achieve an agreement. Many of these groups remained wary of reform efforts and did not share all the goals of Project XL.

MPCA, 3M, and PPC had, with difficulty, created what management scholars might call a "virtual organization" wherein they could work together (Townsend et al. 1998). However, no similar structure existed between the local entities and the regional and national bodies, including environmental groups, whose cooperation was needed for an agreement to gain acceptance. This problem was exacerbated by the fact that the Minnesota project raised issues for which EPA was unprepared. Under these circumstances, EPA's uncertainty about how to respond to conceptual problems that had not been fully anticipated when they designed Project XL became a serious barrier to reaching an agreement.

EPA's approach to these problems, such as how it should handle a facility with many grandfathered emissions sources and extensive voluntary emissions reductions, was not that of a joint problem-solver on a team. Rather, the agency bargained in a time-consuming negotiation with 3M to obtain the best possible deal, meaning the greatest direct environmental benefits. Given the complex issues at hand, this type of bargaining was not conducive to the rapid and effective completion of an agreement. Years of adversarial relations had given rise to a culture within EPA of confrontation and hard bargaining. Though initially 3M did not take this approach, as the negotiations proceeded, both EPA and 3M started to take calculated positions with the full expectation that they would have to relent and make concessions.[32] In the end, this was not conducive to a deal.

Notes

[1] At the North Plant (which had produced magnetic audiovisual tape), 3M was installing new adhesive tape lines. Two of them would rely on a pollution prevention process—reactive coating technology, or hot-melt systems, in which emissions of volatile organic compounds would be negligible because "essentially all the organic raw materials used were locked up on the web as part of the product, rather than being emitted to the atmosphere or requiring the use of energy consuming emission control equipment." Another line would rely on total enclosure and thermal oxidization. It was expected to achieve a control efficiency of 96% or better, a level of control superior to "otherwise applicable requirements" such as best available control technology or New Source Performance Standards. MPCA had granted 3M a permit, which stipulated that the new lines would release no more than 40 tons per year of volatile organic compounds. 3M also planned to convert a magnetic media coater at the North Plant to adhesive tape that would be used in making a pharmaceutical product. Because the solvent-recovery system at the North Plant was not compatible with solvents used in sticky tape production, 3M would totally enclose the new coater and build a thermal oxidizer to achieve control efficiencies of 95% or better (See 3M 1996b).

[2] Letter from C. Williams, commissioner of MPCA, to President Bill Clinton to express disappointment with EPA response to 3M–Hutchinson, July 5, 1996; also see letter from D. Minge, member of U.S. Congress, to C. Browner, adminstrator of EPA, expressing serious disappointment, July 23, 1996.

[3] MPCA, positive comments on the XL Permit for 3M–Hutchinson from Minnesota Environmental Coalition of Labor and Industry; Duluth, Minnesota, Chamber of Commerce; Saint Paul; members of the Hutchinson community; and senior environmental scientist Montgomery Watson, June 14–July 1, 1996.

[4] Twenty-seven people attended, including representatives from various offices of EPA Region 5 and headquarters, 3M, MPCA, and MPCA's Pilot Project Committee (PPC) stakeholder group. The decision to let PPC members come to the meeting was made reluctantly at the last minute. See summary written on August 1, 1996, by A. Ronchak, Project XL coordinator at MPCA, of a meeting of 3M, MPCA, and PPC, on July 30, 1996, at Holiday Inn, Minneapolis–Saint Paul International Airport; also notes by Pilot Project Committee of the same meeting.

[5] The PPC member held that a facility could demonstrate SEP in many ways. The caps in the proposed permit were only one element. The determination of SEP should be based on the average of a company's emissions over a number of years and on how that average improves over time. This average should be compared with what other companies were doing and what was technologically feasible. The assessment of environmental benefits should also take into account factors such as pollution prevention and accountability to stakeholders. People in the community should act as watchdogs, publicizing how well a facility was doing in comparison with its historic performance.

[6] Summary by A. Ronchak cited in Note 4.

[7] Letter from D. Wefring, environmental regulatory specialist at 3M, to A. Ronchak, Project XL coordinator at MPCA, Aug. 7, 1996; letter from J. Kessler, director of regulatory reform at EPA's Office of Policy, Planning, and Evaluation (OPPE), to D. Sonstegard, head of the Environmental Technology and Safety Services Division at

3M, Aug. 16, 1996; letter from D. Gardiner, assistant administrator of EPA's OPPE, to P. Larson, commissioner of MPCA, on how implementing through a site-specific rule would, in effect, replace the rulebook, Aug. 27, 1996; letter from A. Ronchak to F. Hansen, deputy administrator of EPA, on specific compromise language to be added to 3M XL permit, Aug. 8, 1996; letter from A. Ronchak to J. Kessler and M. Martin, innovation coordinator of EPA Region 5, on initial response to EPA counterproposal, Aug. 8, 1996; and A. Ronchak, 3M XL Pilot History, July 30–Dec. 24, 1996.

[8]Letter from D. Wefring, environmental regulatory specialist at 3M, to A. Ronchak, Project XL coordinator at MPCA, summarizing 3M's company objections, August 7, 1996.

[9]Letter from A. Ronchak to F. Hansen, Aug. 8, 1996, cited in Note 7. Ronchak proposed new language for the XL permit, arrived at "after numerous hours of negotiations between MPCA and 3M."

[10]Letter in reply from D. Gardiner to D. Wefring and A. Ronchak, Aug. 8, 1996.

[11]Letter from D. Sonstegard to P. Larson, Aug. 13, 1996.

[12]MPCA passed Sonstegard's offer to Gardiner (letter from D. Sonstegard to P. Larson on resolving remaining substantive issues, SEP, and legal mechanism, Aug. 13, 1996). For the agency, 3M's proposal was just a starting point. EPA wanted more precise language, it said in a response to Sonstegard on Aug. 16, 1996.

[13]MPCA had hoped that Project XL would help it develop three essential tools to help the agency pursue sustainable environmental improvements for Minnesota. These were: first, flexibility for MPCA staff to use innovative approaches to reducing environmental risks; second, a regulatory system that rewards exemplary environmental performance; and, third, alliances among "government, companies, and communities to decide together how best to achieve environmental, community, and economic goals" (letter from P. Larson to C. Browner, administrator of EPA, on never envisioning upfront guarantees for SEP, Aug. 27, 1996).

[14]Letter from D. Sonstegard to D. Gardiner, F. Hansen, D. Ullrich (deputy administrator for EPA Region 5), and P. Larson, on business considerations that 3M must move forward immediately to obtain traditional permits and shelve the Project XL experiment, Sept. 5, 1996.

[15]Ibid.

[16]Letter from P. Larson to C. Browner on how an XL pilot ensures less risk to human health and the environment than current regulatory system, Sept. 17, 1996; letter from PPC to C. Browner on the ways that the Hutchinson permit can provide for SEP, Sept. 19, 1996; A. Ronchak, 3M XL Pilot History, July 30–Dec. 24, 1996.

[17]They were met there by Peder Larson, Andy Ronchak, Dave Sonstegard, the principal 3M–Hutchinson Project XL team members, a Hutchinson plant manager, and one of the facility's leading environmental engineers.

[18]Notes by J. Kessler, "Review of the history of interaction between EPA, MPCA, and 3M on the issue of superior environmental performance and equivalence baselines for the 3M–Hutchinson XL project," on how they can move forward if they reach agreement on treatment of pre-XL voluntary controls in a periodic comparison of actual emissions against what actual emissions would have been absent XL, 1996. Letter from J. Kessler, "RCRA Air Issues for 3M Hutchinson Project XL," to T. Zosel, manager of environmental initiatives at 3M, and A. Ronchak, on what would have

occurred as part of the comparable actions test, October 1996. Letter from J. Kessler, "Proposed Averaging System for Magnetic Tape MACT," to T. Zosel and A. Ronchak, Oct. 18, 1996.

[19]Letter from D. Gardiner to D. Sonstegard, offering to develop an "averaging approach" to MACT as well as seeking assurance that performance under the agreement would be judged relative to the baseline of actual rather than allowable emissions, Sept. 27, 1996.

[20]To our knowledge, during these discussions, the appropriate levels of control efficiencies to assume for the unregulated units in order to calculate baseline emissions levels were never seriously considered. These efficiencies would naturally fluctuate, and some highly efficient units in particular might degrade slightly over time. Even a 1-percentage-point reduction in control efficiency could result in significant emissions increases.

[21]Letter from D. Gardiner to D. Sonstegard, Sept. 27, 1996, cited in note 19.

[22]Letter from A. Ronchak to J. Kessler, on "Draft Hutch XL History and Summary of Changes," Oct. 7, 1996; letter from A. Ronchak to J. Kessler, on "Revised Draft Hutch XL History and Summary of Changes," Oct. 8, 1996; letter from T. Zosel to J. Kessler, on how XL is likely to fall short of original promise, Oct. 21, 1996.

[23]Notes by A. Ronchak on attendance at EPA–3M–MPCA meeting in Research Triangle Park, North Carolina, on Oct. 28 and 29, 1996; summary by A. Ronchak of outstanding issues of that meeting, Oct. 28 and 29, 1996.

[24]Letter from A. Ronchak, "Draft of MPCA Assigned Issues," to T. Zosel and J. Kessler, on Title 5, inter-HAP trading, soft-landing, etc., Oct. 18, 1996; Letter from P. Bartz, air permit engineer at MPCA, "MPCA Assigned Items," to J. Kessler, on HAP trading, comparable actions, new, modified, and existing sources, voluntary reductions, netting, etc., Oct. 21, 1996.

[25]3M would be in compliance with the new MACT standards as long as total HAP emissions coming from audiovisual tape production were less than would be allowed if MACT had been applied to all of the HAP sources regulated by the new standard. EPA went on to amend its MACT rule for magnetic tape production in 1999 (*Federal Register* 64: 17549, April 9, 1999), offering the entire industry 25 options for under-controlling various tanks and compounding vessels in exchange for enhanced control of the tape production lines.

[26]The CAT still left 3M with a great deal of operating flexibility. It did not ask 3M–Hutchinson to exceed regulatory standards by any finite amount—for example, 10% below regulatory requirements—as had been suggested earlier. Nor did it guarantee lower emissions per unit of production, because voluntary controls, under a possible option, could degrade. Substandard controls on new units also were permitted, provided the excess emissions were compensated for elsewhere.

[27]Based on 1996–1997 interviews with EPA staff. See the Acknowledgements for a description of these interviews.

[28]Nor was it entirely clear just how practical the CAT would have been. For example, consider the two new production units installed at the North Plant, which produced adhesive tape products using 3M's new reactive coating technology by which essentially no VOCs or HAPs were emitted. Under the CAT, how do the plant engineers calculate the emissions reduction credits these two pollution prevention units would have earned? What comparable production lines should they use to establish a

VOC emissions baseline? Do such comparable lines even exist? This was only the beginning of the changes at Hutchinson; problems with applying the CAT were likely to multiply.

[29]Based on 1996–1997 interviews with 3M staff. See the Acknowledgements for a description of these interviews.

[30]The proposed XL permit and FPA had the potential to build pollution prevention goals, objectives, and measurements into the ongoing management of the facility because it required 3M to develop a sophisticated environmental management system, which would have played a key role in ensuring compliance. The system included a commitment to establish targets and objectives to reduce wastes and to meet or exceed environmental performance expectations. The agreement committed 3M to report the planning and inception of projects that would have an environmental impact, including whether a project would have source reduction or other effects. 3M had to include projects that were rejected with a brief outline of the reasons that they were not feasible. This reporting requirement strengthened the hands of the engineers and production staff proposing pollution prevention solutions. Especially in cases when the rate of return was only marginally below that of proposals competing for a share of the company's capital budget, this provision would have had the effect of adding additional impetus to a favorable pollution prevention decisionmaking process.

Additional features of the proposed permit and FPA also would have encouraged pollution prevention. By setting broad performance goals rather than specific technology-based or other similar standards, the proposed permit was designed to provide the flexibility to consider a wide range of environmental strategies. By setting overall emissions caps that required SEP over time but allowed flexibility for experimentation with innovative approaches, the permit encouraged pollution prevention. By consolidating, simplifying, and eliminating unnecessary regulatory reporting and other requirements, the proposed permit freed time on the part of 3M staff to develop innovative pollution prevention projects. The proposed 3M XL permit and FPA were designed to allow for much greater public understanding, information, and participation in the regulatory process. They required 3M and MPCA to work closely with both 3M's local stakeholder group in Hutchinson and MPCA's stakeholder group, the PPC, throughout the permitting process from design and implementation to evaluation. Appropriate environmental performance data for the Hutchinson facility would be disseminated on the Internet, and this public disclosure would encourage pollution prevention.

[31]Not all XL team members at EPA were of one mind, however. Some EPA staff members were thinking about systemic changes but also recognized barriers to achieving them. Several staff members pointed out that the 3M proposal, by necessity, had to refer to the existing system as its basis for comparison. Some of the agency's staff were concerned that any real test of the existing system would likely violate aspects of the Clean Air Act. Some expressed the view that XL had a design flaw in that it required fitting projects into the existing legal structure. Without new laws, EPA was handcuffed into doing microprojects, not macroprojects that could change the paradigm (EPA, XLpalooza, meeting summary, minutes, and overheads, College of Insurance, New York, New York, Sept. 11–12, 1996). Without legislation that would allow EPA to conduct pilots that were in apparent violation of existing

laws, the agency could not move toward a new approach to environmental management. Many at EPA, however, disagreed with that view. Although some of the agency's staff believed that XL projects were supposed to be experiments and pilots, they realized that EPA was not explicit enough about what the projects were testing and how to assess their success (1996–1997 interviews with EPA staff; see the Acknowledgements for a description of these interviews). The uniqueness of the companies and their situations meant that it was not easy to transfer specific projects into more general policy initiatives.

[32]The literature on negotiation suggests that parties in a negotiation can take broad classes of actions (Odell 1999). If they follow a *distributive* strategy, they insist on an agreement under which they gain at the other parties' expense. If they adopt distributive strategies, they are likely to divide the gain there is to achieve. Under an *integrative* strategy, they negotiate toward an agreement that makes all sides better off. Experiments (Ross and Stillinger 1991; Kwon and Weingart 2000) suggest that it is counterproductive to start with an integrative strategy. In theory, if the first actor adopts an integrative strategy, then the second actor should reciprocate and the pie should expand. In reality, the second actor typically counters with a distributive strategy. It reasons that the concessions the first actor makes are trivial and have little value. The first actor—offended that its good intentions are not taken seriously—backs off, and the chances for deadlock go up. It is our view that Project XL was set up in this way because business, local stakeholders, and the state government in the 3M–Hutchinson case came to EPA headquarters with integrative solutions, which headquarters tended to discount.

8

Intel, Merck, and Weyerhaeuser: Three XL Projects That Gained Approval

We now turn to issues of substance and process in other Project XL negotiations that were more successful than the 3M Company case. Why were Intel Corporation, Merck & Company, and Weyerhaeuser Company able to initiate negotiations and reach agreements with the U.S. Environmental Protection Agency (EPA), whereas 3M was not? While a Project XL proposal was being designed in Minnesota, these three projects, bearing a number of important similarities to the 3M pilot, reached successful conclusions—if by success we mean reaching an agreement with EPA and being implemented. To better understand the pitfalls that beset the collaborative effort in Minnesota, we turn to an examination of the substantive and process issues in these other cases. We start by discussing their similarities to the 3M project, then argue that their less experimental nature was one of the main reasons it was easier for them to succeed.

Intel, Weyerhaeuser, and Merck: Similarities to 3M

During 1996 and 1997, seven Project XL final project agreements (FPAs) were signed. Of the first seven XL pilots signed in these years, the Intel, Weyerhaeuser, and Merck projects were most comparable to the 3M pilot (see Table 8-1).[1]

All four companies were large, publicly traded corporations known for innovation in both their business and environmental practices. Intel, Merck,

Table 8-1. A Description of the Four Cases

Site	Type of pollution	Agreement
3M Tape Manufacturing Facility Hutchinson, Minnesota, an industrial town about 60 miles west of the Twin Cities; plant is in an attainment area	Air emissions: 2,300 tons per year (tpy) of volatile organic compounds (VOCs); 1,100 tpy of hazardous air pollutants (HAPs) Hazardous waste tanks Minor effluent discharges	Focus on VOCs and HAPs with facilitywide caps; best available control technology on all major sources of pollution prevention reporting; advanced environmental management system; no agreement
Intel Pentium Microprocessor Fabrication Facility Chandler, Arizona (in Phoenix metropolitan area), beset by smog and water shortages; plant is in a nonattainment area	Air emissions: below minor source limits for VOCs and criteria pollutants	Sitewide caps on criteria pollutants and VOCs—all plants considered a single minor source; individual caps on certain HAPs; water treatment and conservation investments; agreement signed Nov. 1996
Weyerhaeuser High-Quality Absorbent Fluff Mill Oglethorpe, Georgia, on the Flint River and in a recreational area near Lake Blackshear in the Flint River water basin; plant is in an attainment area	Effluent discharges into river Air emissions: about 750 tpy of VOCs—mostly non-HAPs Criteria pollutants: roughly a quarter of allowed levels Solid waste (sludge), on-site disposal	Focus on reducing plant effluent discharges into river; dual air emissions caps on VOCs and criteria pollutants; agreement signed Jan. 1997
Merck Pharmaceutical Production Site Elkton, Virginia, adjacent to Shenandoah National Forest in an attainment area but subject to special air pollution regulations as a Class I region	Criteria air pollutant emissions: about 1,100 tpy from coal-fired boilers; area nitrogen oxides limited; about 400 tpy of VOCs	Focus on reducing nitrogen oxides and sulfur dioxide by converting boilers to natural gas fuel; plantwide caps on criteria pollutants; agreement signed Dec. 1997

and 3M were market leaders in their industries and introduced new products frequently. Because they were under intense pressure to stay ahead of the competition, getting these new products to the market quickly was a critical part of their business models. Even Weyerhaeuser's relatively stable Flint River pulp manufacturing facility, producing highly absorbent bleached fluff material, had to be able to respond quickly to changes in demand for various kinds of fluff (e.g., super-white versus a less bright white). For each of these companies, environmental regulatory requirements, especially pertaining to air emissions, were a barrier to rapid responses to changing worldwide market conditions and to introductions of profitable new products.

At these four facilities seeking XL permits, production and environmental engineers followed a process of continuous improvement, looking for better ways to manufacture products and improve quality and for ways to do so with fewer harmful effects on public health and the environment. Their efforts, as well as their introduction of new products, led to frequent adjustments and changes to equipment and materials used for production— adjustments and changes that all too often triggered costly, time-consuming air permit reviews and analyses. The four companies sought the flexibility to implement improvements and process or production changes without burdensome prechange permit reviews, which often simply resulted in verifying that the facility could make such changes without requiring a formal application to the appropriate government agency for change in an existing permit.

To achieve the needed flexibility, all four XL projects included air pollutant emissions caps to allow changes at the facilities without regulatory reviews as long as emissions remained below the caps. These four XL projects were also similar in that negotiating an agreement had high transaction costs (Blackman and Mazurek 1999). Challenging issues surfaced in all four projects, and concessions from all sides had to be negotiated. EPA had hoped that these projects would move from initial selection to implementation within six months. However, Intel, Merck, and Weyerhaeuser each took more than 11 months to negotiate a permit and FPA.[2]

Intel's Ocotillo Campus, Chandler, Arizona

Intel's Ocotillo Campus is at the southern end of Chandler, Arizona, a suburb of about 90,000 people 15 miles southeast of Phoenix. Unlike the 3M, Merck, and Weyerhaeuser sites, the Intel site was in a nonattainment area. The entire basin was beset by smog problems caused primarily by truck and automobile traffic combined with the atmospheric conditions caused by the climate and the mountains to the north of the metropolitan area. The region's limited water supply was another public concern.

Substance

The Intel project, unlike the Merck, 3M, and Weyerhaeuser projects, concerned new facilities, not existing plants with an operating history. In addition, the XL permit and FPA applied to the entire 720-acre site and not to just a single facility built on the site. The facilities planned for the Ocotillo campus were designed to manufacture advanced Pentium microprocessors. Because technological change in the industry was rapid, frequent changes in equipment and in process chemicals were needed (Mohin 1997). Because the facilities were new, there were no well-established baselines from which to measure environmental performance.

Intel's Ocotillo Campus could accommodate at least two major semiconductor fabricating plants (known as *fabs*), and at the time Intel submitted its XL proposal for the site, the first facility on the site, a chip fabrication unit, known as Fab 12, was still under construction. The cost of construction was projected at $1.3 billion, and the unit was expected to employ more than 2,000 skilled workers.

Regulatory Flexibility. Intel's corporate policy was to use "design for the environment" to improve the environmental performance of each generation of production technology and to ensure that all its new facilities would limit their air pollutant emissions so as to operate as minor sources as defined by the Clean Air Act and, therefore, be subject to less-restrictive environmental regulations.[3] Thus, the new plant had been designed to be a minor source and was expected to emit 40% fewer volatile organic compounds (VOCs) than had been emitted by the previous generation of plants (Mohin 1997). However, under most permitting structures, each manufacturing change was subject to some degree of public review (e.g., a 30- or 60-day comment period and public hearing). Intel proposed, as a solution worth testing through Project XL, the preapproval of all these manufacturing changes under appropriately established sitewide air pollutant emissions caps. The caps, therefore, included emissions not only for the newly built Fab 12 but also for any additional manufacturing capacity added to the Ocotillo Campus, a unique feature of this project.

The company also sought an integrated environmental master plan that brought many environmental requirements into a single document and under a single jurisdiction, preferably the Arizona Department of Environmental Quality (AZDEQ). AZDEQ would replace five separate authorities with some jurisdiction over the Ocotillo plants: the EPA Region 9 office in San Francisco, the City of Chandler, the city's fire marshal, Maricopa County, and AZDEQ itself (NAPA 1997). During the course of the negotiations, however, Intel was unable to reach an agreement to consolidate regulatory authority in one agency. Maricopa County insisted on maintaining its jurisdiction over air permitting. The City of Chandler would maintain authority over water treatment issues, and AZDEQ would act as a coordinating body.

The regulatory flexibility Intel received concerned air permit reviews and modifications.[4] The county's air regulations for minor sources were more stringent than were the federal standards for minor sources in a nonattainment area: Maricopa County required the best available control technology (BACT) for facilities that emitted pollutants at a level above an intermediate threshold but still below the federal threshold for requiring BACT. The XL permit did not eliminate this more stringent county requirement. The flexibility gained by Intel did allow the company to install new equipment or process changes without regulatory delays. Although Maricopa County engineers would review these changes for adverse environmental impact, Intel could be assured that production would not be interrupted as long the facility remained in compliance with its XL permit (Jo Crumbaker, Maricopa County Environmental Services Department, private communication, October 2000).

Environmental Benefits. The nonattainment pollutants for the area were ozone (VOCs and nitrogen oxides together form ozone in the presence of sunlight), carbon monoxide, and particulate matter (known as PM_{10} when the matter is 10 microns or less in diameter). Because the facility had low-nitrogen-oxide boilers that burned natural gas as their primary fuel, the direct emissions of these pollutants were low. Hazardous air pollutants (HAPs) and VOCs were of greater concern. Intel had an ongoing program to reduce the need for such chemicals and to control the sources of these compounds. Because Ocotillo had no operating history from which to derive a baseline for determining superior environmental performance (SEP), Intel and the other project stakeholders decided to use the regulatory requirements for minor sources as the baseline. Intel management was determined not to challenge existing regulations. Thus, in contrast to the other projects, Chandler did not require a site-specific rule.

Intel's XL project contained several environmental benefits. The entire site that would in time contain two or more fabs was treated as a single "minor" pollution source and therefore would emit less than half of the regulated air pollutants the federal regulations allowed. The company agreed to limit emissions of each HAP so as not to exceed the Arizona Ambient Air Quality Guidelines, which were risk-based safety guidelines for approximately 400 chemicals and provided an environmental benefit not only to the local residents but also to the fab workers.

Intel also was committed to funding a reverse-osmosis water treatment plant on site. This plant would treat more than half of the process water, meet drinking water standards, and return water to local aquifers or use it for irrigation. This $25 million water treatment plant would be almost entirely funded by the company and would be operated and owned by the City of Chandler. Intel donated to the city the land upon which the plant was to be built and would pay almost all the operating costs—$500,000 to $1 million a year. The company also agreed to conserve water use on site in other ways and to go beyond compliance with storm water regulations by pumping storm water into a retention basin, thereby decreasing the impact on groundwater.

The FPA included additional environmental commitments by Intel: to increase the recycling of solid waste over time by 60% and of chemical waste over time by 70%, to make integrated reports of the site's environmental performance more easily available to the public, and to promote and encourage carpooling among the workers at the Ocotillo Campus. The company also promised that all buildings on the campus would be set back at least 1,000 feet from the nearest residences. Finally, a single consolidated emergency response plan would be developed with the help of the Chandler Fire Department to improve responses.

Despite these innovative attributes, the Intel XL proposal was not groundbreaking. As the National Academy of Public Administration reported, "In contrast with some other XL proposals and the original public understanding of the program, Intel's requests were modest" (1997, 79). Yet, despite the modest amount of regulatory flexibility Intel wanted and the high level of SEP ensured, it took almost a year before a final agreement was reached.

Superior Environmental Performance. Table 8-2 summarizes the air emissions caps established in the FPA and five-year XL permit. The best comparison is with the minor-source thresholds for two facilities, not just for Fab 12.[5] The permit limits for sulfur dioxide and PM_{10} were small fractions of conventional minor-source limits, and the carbon monoxide cap was only 25% of the conventional permit limit. The VOC cap was 40%, and the nitrogen oxide cap was half of the minor-source limits.

The FPA also included pollutant emissions per unit of production as parameters. To define the unit of production, the concept of *production unit factors* (PUF) was introduced. Rather than simply take the total area of silicon microprocessors produced, the rapid increase in the number of transistors contained in a given area of silicon was taken into account.[6] Tracking the emissions of a pollutant divided by PUF would help verify that emissions reductions were not simply due to a drop in output and, conversely, that emissions increases could be only attributable to production increases (Intel Project XL Progress Report, December 1999).

Clearly, the XL permit set enforceable air emissions limits that were considerably better than what the "otherwise applicable regulations" required. EPA could have chosen to raise the question of whether Intel's environmen-

Table 8-2. Air Pollutant Emissions Caps on Intel's XL Permit for Its Ocotillo Campus (tons per year)

Pollutant	Conventional permit limit for a facility designated a minor source	Conventional permit limit for two minor-source plants on campus	Project XL permit for campus sitewide caps
Carbon monoxide	<100	<200	49
Nitrogen oxides	<50	<100	49
Sulfur dioxide	<250	<500	5
PM$_{10}$	<70	<140	5
Volatile organic compounds	<50	<100	40
Hazardous air pollutants (HAPs)	<25 aggregate; 10 for any individual HAP	<50 aggregate; 20 for any individual HAP	20 aggregate: 10 organic HAPs; 10 inorganic HAPs (4 phosphine; and 9 sulfuric acid included in inorganic HAPs aggregate)

Note: PM$_{10}$ is particulate matter 10 microns or less in diameter.

Source: Intel Permit and Final Project Agreement

tal performance at the Ocotillo Campus would have been any different under conventional regulation. Perhaps recognizing that the question was unanswerable, the agency chose not to raise it. Instead, EPA argued in defense of its approval of the project that these proposed emissions levels were no worse, and possibly better, than those at other comparable semiconductor manufacturing plants in the United States. Nonetheless, approval of this XL project did not proceed without controversy.

Process

To begin the project, Intel formed a 15-member stakeholder group whose task was to develop and approve the FPA. Members of this team included representatives from Intel and the public agencies involved: the Maricopa County Environmental Services Department, AZDEQ, EPA Region 9, and Chandler's water and fire departments. Completing the stakeholder group were five citizens representing the public: a school board member, an environmental and land-use consultant, a software engineer and community activist, and the director of the Gila River Indian community's Department of Environmental Quality. EPA headquarters was represented at the first meeting and participated by telephone during the "core" meetings.

Intel hired the Denver Research Group to facilitate the meetings, and Intel and EPA provided support staff, including experts to help with legal and engineering issues. As in the other early XL projects, no funds were available for the citizens in the Intel stakeholder group to hire their own consultants to provide an independent assessment of the many technical issues that came up. An Intel employee assigned to do outreach contacted several nongovernmental organizations (NGOs), most of which made some effort to keep abreast of the project but did not follow it closely. An environmental organization called Don't Waste Arizona actively participated and supported the final agreement.

The company received approval in November 1995 to develop an XL project, and the FPA was signed November 19, 1996. The stakeholder group began meeting in late January, breaking up into working teams to tackle particular parts of the FPA. The group met about two to three times a week for nine months, for a total of about 100 meetings, which frequently went from 6:00 p.m. until midnight. Initially, a lot of time was devoted to educating the citizen representatives about the many technical issues involved. Eight of the meetings were designed for public participation, with efforts made to notify and encourage citizens to come. According to the National Academy of Public Administration report, the entire "endeavor was careful, complete—and stressful" (NAPA 1997, 90).

Although some of the stakeholders frequently raised questions and challenged the majority (most notably the community activist, who was a vocal critic of the FPA throughout the process, although he nevertheless signed it), the group engaged more often than not in cooperative problem solving. The presence of representatives from EPA's regional office was a key to the project's success. Region 9 representatives attended many of the meetings, spending thousands of dollars on travel expenses.

The Negotiations. The air permit was complex and difficult to complete. One stakeholder member estimated that about 14 different versions of the air permit had been considered before it even reached the public review stage. According to an Intel participant, the air permit was so frustrating and difficult that at one point he thought Intel should consider withdrawing from the project. Intel management originally had approved a six-month project.

Eventually, the stakeholders arrived at a consensus for a draft FPA and XL permit, only to discover that their deliberations were not at an end. Intel presented the outlines of its XL project to NGOs and to staff from EPA's Office of Policy, Planning, and Evaluation (OPPE) at a May 15, 1996, meeting in Washington, D.C. Many of the questions that had come up at stakeholder meetings were raised again during the question period. What was the proper baseline to use? What about trading more toxic HAPs for less toxic ones under the cap? Is this SEP? In follow-up comments in a July 3, 1996, letter, the Natural Resources Defense Council (NRDC) presented its many objections to the project (to the Merck and 3M projects as well) in some detail. Arguing it did not have the resources to participate in all the ongoing XL projects and complaining that the Intel group kept its drafts confidential, NRDC took its case directly to top officials at EPA headquarters and, according to the NAPA case study, to the top managers at Region 9. Final approval was delayed.

Many of the participating stakeholders were upset with this end run around their negotiating process. The EPA Region 9 participants, who thought they had been given the authority to make binding decisions (presumably after consultation with the regional office and headquarters), found themselves undercut. Several NGOs and one of the stakeholder members were working against the agreement outside of the stakeholder process. The community activist in the stakeholder group joined the effort to get modifications in the agreement. Though the questions raised did not result in any major changes in the FPA or permit, Maricopa County did get public accountability commitments in the FPA inserted into the permit as enforceable requirements. EPA finally signed off on the agreement, and the Maricopa County Environmental Services Department issued a permit to Intel.[7]

In comparison with the 3M, Merck, and Weyerhaeuser XL projects, Intel's proposal was a modest one that did not require a site-specific rule from EPA. The company agreed to an air permit with, on balance, much more stringent emissions limits than what the alternative, a regular Maricopa County permit, would have required. Then why was this project so controversial, and why did it take so long to reach a final agreement?

One participant in the stakeholder group observed that the greatest obstacle came from the environmental advocacy organizations. She singled out the Silicon Valley Toxics Coalition, which (according to the NAPA report), "spent many years fighting Intel and other California-based computer companies over past ground water contamination" (NAPA 1997, 92). This viewpoint is reinforced by the fact that objections to the agreement did not end with the FPA's signing and XL permit issuance. A letter of protest signed by more than 100 individuals and representatives of various environmental, community, and labor groups was issued on the day of the FPA signing. The protesters warned of employees' and residents' increased exposure to toxic chemicals. NRDC's press release complained that the project should have achieved more for the environment.

Intel carried out an ambitious negotiating process with its stakeholder group, without which, most probably, agreement would not have been possible. Nevertheless, this negotiating process did not insulate Intel from strong resistance from others outside the process and even from someone who was part of the process. The NAPA report put it well:

> The Intel case makes evident how difficult it is to prove to all stakeholders that a "trade" of increased flexibility for a commitment to better performance produces net social and environmental benefits. Science and economics may not resolve those issues, because technical uncertainties and differing values leave the process of making a deal or defining "superior environmental performance" prone to technical, political, and legal disputes. (NAPA 1997, 75)

Fab 12 began initial operations in the summer of 1996 but did not fully ramp up production until 1998.[8] A small expansion of the plant was made soon after, and a second large plant, Fab 22, began operating in 2001.

Weyerhaeuser's Flint River Plant, Oglethorpe, Georgia

Weyerhaeuser's Flint River Plant manufactures high-quality absorbent fluff. The facility, located in an attainment area, is on the Flint River near Oglethorpe, Georgia, a town of about 13,000 people roughly 100 miles south of Atlanta. In 1996, the plant employed about 500 people with an annual payroll of more than $20 million that contributed an estimated $75 million a year to the Georgia economy. These economic factors were made known to EPA when Weyerhaeuser submitted its request for an XL permit. The plant is situated in the Lake Blackshear watershed, an important ecological and recreational resource. The Flint River flows through Lake Blackshear, 20 miles to the south, then southwest and joins the Chattahoochee River and Lake Seminole at the Florida border. The combined waters then continue south as the Apalachicola River and enter the Gulf of Mexico about 80 miles away.

The river is both a source for the plant's process water and a sink for its treated process effluents. As the document *The Weyerhaeuser Company: Frequently Asked Questions about the XL Project* (www.epa.gov/projectxl/weyer/011797_f.htm, accessed June 18, 2002; hereafter, *Frequently Asked Questions*) noted:

> Local community members of the Lake Blackshear Watershed Association have a history of concern for, and protection of, the River. In the past, concerns have been raised about the health of turtles and other wildlife in the river due to pollution from pulp and paper mills on the Flint River. Protection and preservation of the river have been shared goals of Weyerhaeuser and the local community for many years.

The Flint River Plant, designed to produce bleached kraft pulp, was constructed on a 2,000-acre site by Procter & Gamble in 1980 but sold immediately afterward to the Weyerhaeuser Company. Weyerhaeuser then proceeded to invest $200 million to create a highly efficient operation (Dawson 1998). Thereafter, Weyerhaeuser continued this commitment to maintain the facility as an industry leader. As the project's FPA said:

> Flint River Operations' environmental performance has been recognized as superior within the bleached Kraft pulping industry. Flint River was the first bleached Kraft pulp mill to employ commercially viable advanced technologies that minimize adverse impacts to the environment. These technologies include oxygen delignification (installed in 1980), 100% chlorine dioxide substitution and bleaching (in 1989) and extensive water conservation practices.

Oxygen delignification is a cutting-edge pollution prevention technology that reduces the amount of bleaching chemicals needed in the wood pulp process. Chlorine dioxide substitution (that is, 100% substitution of chlorine dioxide in place of elemental chlorine gas) has been instituted at Flint River as one of several methods used to reduce the formation of unwanted chlorinated organic compounds during the pulping process. Water conservation measures reduce raw water costs and the volume of wastewater for treatment and discharge. These investments were forerunners of a process of continuous improvement called Minimum Impact Manufacturing (MIM), which Weyerhaeuser introduced in 1992.

Flint River Plant staff sought to manage the facility's raw material and other resources to achieve continuous improvements in water, air, and solid waste discharges. MIM was in place prior to Weyerhaeuser's application for an XL permit and would continue to be in place after the permit. Consequently, many of the improvements in environmental quality Weyerhaeuser would make with its XL permit were improvements it was planning to make before the permit was signed.

The Flint River Plant Operations

The Flint River facility manufactures highly absorbent, white (i.e., bleached) fluff pulp for use in such products as disposable diapers (such as Pampers— Procter & Gamble is a major customer). Although the plant's total output was fairly constant, averaging about 320,000 tons a year, the product did vary in the degree of whiteness or, in other words, the amount of bleaching. Demand for white and super-white versions of the fluff fluctuated, and the company had to be able to respond quickly to these market changes (Gary Risner, Weyerhaeuser Company, private communication, September 2000).

The Flint River Plant's main problem with the environmental regulations governing its operations concerned the requirement in the Clean Air Act to perform prevention of significant deterioration (PSD) reviews. For example, if customer demand for super-white absorbent fluff were to increase, Flint River would have to adjust its production process to increase the amount of bleaching. This modification would require an increase in the output of one or more of its boilers, an increase allowed by current regulations. Nevertheless, if as a result of this process change, projections showed that total emissions of any pollutant regulated under PSD would be increased above the PSD threshold, Flint River engineers would be required to undergo a PSD review, which would take plant staff on average six months to complete. Because the pollutant emissions increase was due to a boiler output increase, which was allowed by regulations, the PSD review would inevitably conclude that there was no need for a permit revision.

Water pollution was the primary environmental concern for the area. Consequently, reducing the pollutants contained in the treated wastewater (effluent) discharged into the river had the highest priority and was the principal, although not exclusive, focus for SEP at the facility. Several quantities or parameters are commonly used to measure the environmental impact effluents have on the waters into which they are discharged. Biological oxygen demand (BOD) is the measurement of the oxygen consumed in an effluent sample by the biological processes breaking down organic matter (what was used was "BOD5," the oxygen consumed in 5 days). Chemical oxygen demand (COD) is the oxygen required to oxidize the organic and inorganic compounds in the effluent. The term total suspended solids (TSS) is self-explanatory, and adsorbable organic halide (AOX) is a measure of the amount of chlorinated organic compounds in the effluent sample. These parameters are defined in terms of weight of pollutant per unit of production, which in this case is air-dried metric ton (ADMT) of fluff. A metric ton is 1,000 kilograms (kg), or 2,205 lb. (1.10 tons). Thus, the methods for measuring SEP in water pollution, the main focus of SEP in the Weyerhaeuser permit, were in emissions per unit of production.

Weyerhaeuser's discharges were well below the discharges at other comparable plants, and BOD, TSS, and AOX baseline discharge rates were less than the new BACT requirements for the industry. A Weyerhaeuser environmental manager estimated that the Flint River Plant might very well have been the best-performing mill of the approximately 100 plants manufacturing fluff in the United States (Gary Risner, Weyerhaeuser Company, private communication, September 2000). Two other related environmental performance parameters were Flint River water usage, with a baseline rate of 11.2 million gallons per day, and "bleach plant flow," effluent from the bleaching part of the production process, with a baseline rate of 20 cubic meters per ADMT of fluff produced. By industry standards, the rates of water usage and bleaching effluent produced per unit of production were both very low.

Operations at the Flint River facility also were responsible for significant air emissions of criteria pollutants and VOCs. The major sources—boilers to produce steam for process heat and to generate electricity, a calciner, and a pulping solution dissolving tank—were responsible for all the plant's sulfur dioxide and nitrogen oxide emissions (in 1995, respectively, 587 and 881 tons per year, or tpy) and a large fraction of the emissions of particulate matter and carbon monoxide. A large part of the total plant releases of VOCs and HAPs (in 1995, respectively, 759 and 534 tpy plantwide) came from operations at the rest of the facility (Weyerhaeuser 1998 XL Progress Report; Dawson 1998).

The process of producing absorbent fluff from timber also yielded considerable amounts of solid waste. In 1995, roughly 0.3 pounds of solid waste were generated for every pound of fluff produced. Very little hazardous waste was generated, however. The waste never left company land and was deposited at a modern landfill on Weyerhaeuser property (Final Project Agreement; Gary Risner, Weyerhaeuser Company, private communication, September 2000).

According to EPA, Flint River's intake of river water, which was about 12 million gallons per day, was less than half of the average water used by other comparable plants (25 million gallons per day). Less water taken in means less effluents out, and the plant was used by EPA as a benchmark for effluent guidelines and "liquor best management practices" (Final Project Agreement; EPA Responds to Comments). Moreover, the facility had embarked on its ambitious six-phase MIM program for continuous improvements in environmental performance and was currently finishing the third phase. Weyerhaeuser environmental managers thought that Project XL would be a good vehicle to continue these efforts (not start new ones), especially if regulatory flexibility could be obtained that would reduce barriers to achieving the goals of the program as well as provide economic benefits to the company.

Substance

The environmental benefits of the project fell into two categories: the short-term, relatively predictable benefits that would derive from the implementation of the fourth and parts of the fifth phases of the MIM program and the longer-term goals of the rest of the fifth phase and the sixth phase. The shorter-term, predictable environmental benefits, most of which became enforceable provisions of Weyerhaeuser's XL permit, included installing a state-of-the-art delignification unit (lower temperature, slower cooking) and other related improvements that would lower effluent discharges, reduce costs, improve the product, and increase output by 2%. Other phase four improvements were to upgrade odor controls, reduce the need for energy steam, and implement the new environmental management system. Included in the short-term category was a phase five program to implement strategies to improve 300,000 acres of Weyerhaeuser's Georgia timberland for wildlife habitat, protect threatened species and conserve other species, reduce soil erosion and improve the management of streams and streamside areas, and improve aesthetics.

The effluent reductions at the Flint River Plant were embodied in enforceable provisions of the XL permit that limited BOD5, TSS, and AOX discharges to 21%, 52%, and 4%, respectively, below the BACT requirements for the industry (see Table 8-3). In addition, as part of the MIM process of continuous improvement, the facility would seek to reduce energy usage (process steam and electricity). Weyerhaeuser pledged that any reduced need for electrical energy (self-generated on site) would result in less boiler use and, therefore, reduced air pollutant emissions. The longer-term elements of the project were nonenforceable goals in the FPA to strive for and be achieved, if possible, by 2006. Using the average rates for the three-year period 1993–1995 as a baseline, these goals—starting with the most ambitious—included reducing effluent from the bleaching plant (containing chlorinated compounds and other pollutants that were the number one concern) by 50%, reducing solid waste per ADMT by 55%, and reducing raw water usage by 9%.

Regulatory Flexibility. Weyerhaeuser sought relief in the XL permit from air pollution PSD requirements, which it found to be a bureaucratic barrier to its MIM program and to a more rapid response to changing customer

Table 8-3. Weyerhaeuser's Effluent Discharges: Permit Limits and Final Project Agreement Goals

Environmental parameter	Prior permit limit	BACT requirement	Baseline (1993–1995 averages)	XL permit limits	Goals for final project agreement
Bleach plant flow	None	None	20 cubic meters per ADMT fluff	None	10 cubic meters per ADMT fluff
Total water usage (millions of gallons per day)	14.34	None	11.18	11.5	10.18
Biological oxygen demand (lb. per ADMT)	5.3	4.83	4.32	3.80 and 4,826 lb. per day[a]	2.5
Total suspended solids (lb. per ADMT)	5.8	8.58	4.65	4.09 and 5,194 lb. per day	2.1
Adsorbable organic halides (kg per ADMT)	None	0.156	0.11	0.15 and 190.5 kg per day	

Note: BACT = best available control technology; ADMT = air-dried metric ton.

[a] These daily XL permit limits are obtained by multiplying the limits established per ADMT by 1,270 ADMT, the maximum daily output of fluff at the plant.

Source: Weyerhaeuser XL Permit and Final Project Agreement.

demands. The company wanted facilitywide air pollutant emissions caps under which the plant could make changes without undergoing time consuming PSD reviews. After negotiations, however, the company agreed to a more complex system of dual caps on six individual pollutants. (See Table 8-4 for the individual pollutants covered and the magnitudes of the caps.) One set of caps was for total emissions coming from the four major emissions units. These sources were already regulated to comply with BACT standards and were allowed to emit pollutants up to their potential to emit. Consequently, the caps on individual pollutants were set by taking actual emissions in 1995 and scaling them up to the level that would result from the facility operating at its maximum output of 1,270 air-dried metric tons a day. This approach was very similar to the approach taken in the 3M case—with one important difference. Under these major-source caps, no flexibility was granted. Any physical or operational modification of these units would have to undergo the normal PSD review (Appendix 6 of the Final Project Agreement).

The second of the dual air emissions caps on the same six pollutants applied to total plantwide air emissions coming from the operations of the rest of the facility (i.e., excluding the plant's major sources). These six caps were set more stringently by taking actual 1995 emissions and adding the PSD significance levels for individual emissions units (that is, 40 tpy of sulfur dioxide and 40 tpy of nitrogen oxides) that triggered PSD reviews (see Table 8-4). Thus, as long as plantwide pollutant emissions remained below these caps, no modifications of any of the applicable emissions units could ever trigger a PSD review.[9]

Weyerhaeuser also obtained regulatory relief in simplified, consolidated biannual reporting, and the possibility after two years that the parties would agree to only annual reporting. In addition, periodic testing of fish samples for dioxin was no longer required, because none had been detected for several years. In addition, under the XL permit, Weyerhaeuser would no longer have to dry out its solid waste, much of which was in the form of one kind of sludge or another (predominantly lime mud), before disposal at its own lined landfill (Weyerhaeuser Final Project Agreement and Appendices). Finally, Flint River received meaningful relief in how it had to comply with the maximum achievable control technology (MACT) standard to control HAPs, which was soon to be promulgated (MACT Cluster rule for the pulp and paper industry). As was noted above, the MACT standard established in the 1990 Amendments to the Clean Air Act was to be set for each manufacturing category and was defined as achieving control efficiency not less than that achieved by the best-performing 12% of existing controlled sources in that category. The company was given a site-specific plantwide emissions rule as an alternative to a source-by-source application of the new regulation. The facility's XL compliance rule allowed it to cap its HAP emissions

Table 8-4. Weyerhaeuser Flint River Plant Air Emissions and XL Permit Caps (tons per year)

Environmental parameter	Allowable plantwide emissions under the previous permit	Actual plantwide emissions in 1995	Emissions caps in XL permit for major emissions units	Emissions caps in XL permit for the rest of facility
Volatile organic compounds	1,945	759	75	703
Carbon monoxide	6,290	1,780	2,170	346
Nitrogen oxides	3,250	881	1,260	40
Sulfur dioxide	2,197	587	839	40
Total reduced sulfur (odor control)	155	43	35	27
Particulate matter	1,472	457	394	195

Source: Weyerhaeuser XL Permit and Final Project Agreement.

facilitywide to no more than the total emissions that would have resulted if all the regulated emissions units were controlled to the MACT standard.

This alternative compliance rule was very similar to the one EPA offered 3M for its magnetic tape production units. Weyerhaeuser received approval for this alternative strategy in mid-1996, several months before EPA offered it to 3M (Appendix 7, Final Project Agreement). Under this alternative rule, the plant received credit for all HAP emissions reductions that were implemented on or after January 1, 1996. This amounted to credit for past good deeds only one year into the past but nevertheless was a more generous credit than offered to 3M. The company had installed earlier, as part of its odor control program, a vent collection system that had the added benefit of controlling HAPs that were mixed in among the collected gases. The XL permit gave the plant credit for these HAP reductions. Because some of these vents were unregulated sources, Weyerhaeuser received some credit for past voluntary controls. Unlike the 3M case, this credit involved HAP emissions reductions of only 40 tpy, just 8% of the HAP reductions to be achieved by the new rule. The MACT bubble and credits for some of its pre-XL investments were an important issue for Weyerhaeuser, saving it an estimated $10–20 million in capital investments that would have been required by direct compliance with the new MACT standard (Gary Risner, Weyerhaeuser Company, private communication, September, 2000; Appendix 7, Final Project Agreement; *Frequently Asked Questions*).

Superior Environmental Performance. Before Project XL, Weyerhaeuser's Flint River Plant had one of the best, if not the best, environmental performance records of any absorbent fluff-producing mill in the country. For the plant to generate under XL an environmental performance superior to an already outstanding performance could have been a difficult hurdle for the company to overcome. In addition, EPA staff members could have argued that Weyerhaeuser, given its MIM program and past record, would have done all these things anyway without any regulatory flexibility (Gary Risner, Weyerhaeuser Company, private communication, September 2000). Instead, EPA defended the environmental benefits of the project in its *Frequently Asked Questions* and in its response to comments made about the proposed FPA. The EPA singled out the voluntary target in the FPA to cut "bleach plant flow" in half by 2006 as the project's greatest environmental benefit. Nearly all of the AOX and TSS came from the bleach plant, as well as two-thirds of the facility's effluent responsible for BOD (see Table 8-3). The EPA also cited the company's voluntary target to reduce water usage by almost one million gallons per day, even before the reductions at the bleach plant took place. This would reduce water intake by possibly another two million gallons per day. Although these goals were not enforceable, only voluntary, and were by no means a certainty, they were ambitious and, if even only partially achieved, would clearly be SEP. They stretched the limits of what was technologically possible and, if proven at Oglethorpe, might be more widely adopted in the industry.

The EPA document also cited Weyerhaeuser's commitment to upgrade the environmental management of the approximately 300,000 acres of timberland in Georgia. EPA thus showed that it was impressed by the comprehensiveness of the Weyerhaeuser accord. This part of the project, moreover, was highly doable and therefore predictable; indeed, it was implemented before the end of 1997 (Weyerhaeuser Project XL Annual Progress Reports). This component of the project's SEP was unrelated to the flexibility granted the plant and seemed to be part of the overall transaction to assure EPA that sufficient SEP was being exchanged for the flexibility EPA was granting. If the bleach plant effluent and water use reductions were ambitious goals for SEP that might not be achieved, the timberland improvements were guaranteed elements of SEP in the project, and it was guarantees that EPA was seeking.

Another important component of the direct environmental benefits that were guaranteed as an enforceable part of the XL permit was the new permit limits set for BOD, TSS, and AOX (see Table 8-3). *Frequently Asked Questions* pointed out that, in addition to these new limits being less than the BACT standards proposed at the time, the BOD limit was 28% below and the TSS 29% below the previous permit limit. In addition, as shown in Table 8-3, both the BOD and the TSS XL permit limits were set below average *actual* emissions during the baseline years. AOX, which can build up in fish tissue, had no limit under the earlier permit. Water usage, about 60% lower than at a typical fluff plant, was also limited as an enforceable element of the XL permit. Although it was set slightly above the baseline, it was set 20% below the previous permit's water use limit. EPA also noted the voluntary targets in the FPA to reduce solid waste generation by half over 10 years and to reduce energy consumption roughly 10% by recovery and reuse of process steam. Weyerhaeuser's permit stretched technological limits and was very comprehensive in nature, but in large part was a continuation of commitments the company already had made before its application for an XL permit.

Although the focus of the Weyerhaeuser XL project was reducing effluent discharges into the Flint River, EPA also defended the air emissions caps as a part of the superior environmental performance contained in the permit and FPA. The four major sources responsible for almost all of the criteria pollutant emissions were already limited to BACT standards. The caps were merely set to the maximum allowed levels under BACT when the plant operated at full capacity. Nevertheless, these caps were set far below the limits allowed in the facility's previous permit and, in any case, no flexibility was granted under these new facilitywide limits. Relief from frequent PSD reviews was provided under the caps on air emissions (predominantly VOCs) by all other sources at the plant.

Unlike in the 3M case, these limits were set very close to actual emissions, especially for VOCs. EPA commented that, in contrast, under the "otherwise applicable regulations," the Flint River facility could have made a series of changes over time that, little by little, increased air emissions to levels well above the XL permit caps but that were each below the PSD significance levels, thereby avoiding PSD reviews. EPA also defended the FPA against criticism that there was no basis to grant credits toward the MACT requirements from voluntary controls on non-MACT sources. The agency admitted that in the short-term the credits could result in less reduction of HAPs than under source-by-source compliance with MACT. But, the agency said, the credit was small (40 tpy out of total reductions of about 500 tpy), and in the long term the overall project would lead to greater HAP reductions than would simple compliance with the MACT standards.[10]

Process

Before proceeding to prepare a formal proposal in 1995, Weyerhaeuser presented the outlines of the project to Jon Kessler of OPPE and others at EPA headquarters in Washington, to the Georgia Environmental Protection Division, and to local stakeholders such as the community's Thought Leaders group. Everyone was encouraging, and Weyerhaeuser proceeded to apply. Its XL proposal was accepted in December 1995 (Gary Risner, Weyerhaeuser Company, private communication, September 2000). In January 1996, Weyerhaeuser formed a stakeholder group to advise on and help develop an FPA. Included in the group were representatives from water, air, and other programs at EPA Region 4 and at the Georgia Environmental Protection Division; the Georgia Pollution Prevention Assistance Division; the Lake Blackshear Watershed Association; the cities of Oglethorpe and Montezuma; Macon County officials; and nonmanagement Flint River Plant employees. Aside from Weyerhaeuser employees, the only nonofficial citizen representatives were from the Lake Blackshear Watershed Association.

The Region 4 participants, according to Gary Risner of Weyerhaeuser, were enthusiastic about Project XL and the opportunity to work "outside the box," and Weyerhaeuser had an excellent working relationship with the Georgia Environmental Protection Division. The stakeholders were able to reach an agreement on the issues involving solid waste and water quality relatively early in the discussions. Issues concerning air pollutant emissions were more difficult to resolve, and air experts at the Georgia Environmental Protection Division and representatives from the Georgia Pollution Prevention Assistance Division had to devote more time to the project. A representative from EPA's Office of Air Quality Planning and Standards also played an important role in forging an agreement, especially concerning the alternative compliance strategy for the new MACT Cluster rule (Dawson 1998;

Gary Risner, Weyerhaeuser Company, private communication, September 2000). EPA noted that NRDC, Environmental Defense, and the Institute for Regulatory Policy also provided input at several stages in the development of the agreement (*Frequently Asked Questions*).

According to Risner, Michele Glenn from Region 4, who became project director, did a "super job" of coordinating the project and getting the different parties to agree. Excellent progress was made, indeed so much progress, that he thought they might have an agreement by May 1996, within the six months originally put forward by EPA. Risner thought the negotiating process was cooperative and parallel rather than linear and back and forth from one party to another. There were few delays in distributing information to the parties, and Weyerhaeuser people had decisionmaking authority delegated to them by top management, who had already signed off on phase four of the company's MIM program. Everyone seemed to be able to come together at the same time and the same place to discuss difficult issues when they arose. A top Project XL manager at EPA Region 4 advised Weyerhaeuser staff early in the process not to be vague about what they wanted, and the staff followed these instructions. Also helpful were both an outline of the project agreed to early in the process and established timelines.

Negotiating with EPA Headquarters. However, the project needed the approval of EPA headquarters. To that end, a meeting was held in the spring of 1996 with EPA people from OPPE, the Office of General Counsel, the Office of Water Quality, and other offices. The meeting did not go well, and afterward members of Weyerhaeuser's XL team were ready to walk away from the project. Instead, they decided to wait and give EPA headquarters and themselves a cooling-off period of a month or so. One possible reason the meeting went so badly was that OPPE staff and perhaps others at headquarters were disappointed that the proposal did not include a goal to create a closed-loop bleach plant operation (i.e., zero bleach plant effluents). They were hoping that Weyerhaeuser, as an environmental performance standard setter, could demonstrate for certain that the closed-loop process worked, which could then be promulgated as a standard throughout the industry (Gary Risner, Weyerhaeuser Company, private communication, September 2000).

EPA did sponsor a meeting on May 15, 1996, facilitated by the Keystone Center, where Weyerhaeuser presented the ideas for its XL project to an audience of NGOs, community organizations, and EPA staff. This meeting helped smooth over some of the tension. During the question and comment period, several participants raised concerns about the lack of compliance mechanisms to enforce the performance of actions under the FPA. To address the reporting and enforcement concerns, Weyerhaeuser stated that it would establish a schedule of semiannual and annual reporting and a rigorous system of monitoring requirements. In addition to enforcement issues,

Weyerhaeuser representatives answered questions concerning topics such as the use of chlorine dioxide in their pulp bleaching process, plans for timberlands management, the use of virgin versus recycled fibers, and the ability of stakeholders in the XL process to evaluate technical issues. Representatives from Environmental Defense and NRDC had been invited to stakeholder meetings early on, and Risner of Weyerhaeuser had two private meetings with David Hawkins of NRDC. Some of these inputs were incorporated into the FPA.

Finally, negotiations resumed with EPA headquarters. But unlike the negotiations in the stakeholder process, these discussions were more linear, more like win–lose negotiations during which EPA took tough positions at first. Progress was made; even so, during the summer of 1996, the Weyerhaeuser XL team was feeling worn out by the continual back and forth negotiations, especially with EPA's Office of General Counsel.

For Weyerhaeuser, what made the deal possible was approval for the facilitywide approach to complying with the new MACT Cluster rule. It had to be granted some credit for its past good deeds for the deal to work. Source-by-source compliance would require capital investments of an estimated $10–20 million with very little environmental benefit to show for it. To obtain what it wanted, the company did make some concessions. It accepted the dual cap structure, met EPA headquarters halfway, and agreed to aim for a 50% reduction in bleach plant flows by 2006 as a voluntary commitment in the FPA. In the end, agreement was reached, but not until Weyerhaeuser staff ended the back-and-forth linear process with headquarters by asserting that it would drop out if one last attempt at agreement failed.

Other factors that contributed to the agreement included the Flint River Plant's position as an environmental leader and the fact that if the project was successful, the whole industry would benefit. Another important factor was that a clear baseline from which to measure environmental performance had been established.

Success in reaching agreement was also helped immeasurably by Weyerhaeuser's having champions for the project within EPA (Weber 1998). Region 4 participants were strong advocates. During the cooling-off period with headquarters, a new administrator was appointed at Region 4. He was invited to visit the Flint River facility to learn about the project. As a result, he became a strong and influential advocate of the XL project and was a major factor in its final approval by EPA headquarters (Gary Risner, Weyerhaeuser Company, private communication, September 2000). The EPA regional office did an effective job in getting support from the various parties and in guiding the process, but Weyerhaeuser also had help from an EPA staff person in the Office of Air Quality Planning and Standards who worked on the air emissions issues, was an enthusiastic supporter, and gained the support of the office's administrator, who helped sell the project. Weyerhaeuser's ability to work constructively with EPA, however difficult its relations with the agency were at times, was quite different from 3M's inability to do so. 3M had good working relationships with state government and worked very closely with the state pollution control agency, a type of relationship that it could not recreate with EPA Region 5 or headquarters.

Unlike Intel, where the local community stakeholders raised several important issues persistently, the community surrounding the Flint River Plant did not seem to be concerned about the project. This situation was certainly due in part to the exemplary environmental performance of the facility after 1980.

Finally, it is worth noting that EPA, in the end, decided to incorporate the alternative compliance strategy it had granted Weyerhaeuser for the new MACT Cluster rule as a permanent part of the rule, offering this option to all others in the industry. Though this action was the kind of change one can imagine emanating from Project XL, given its goals, it was puzzling that EPA took this action so quickly, before testing the idea at Flint River and in light of the difficulties raised by EPA headquarters when it was first presented with the draft FPA.

In the five years of implementation of the Weyerhaeuser project, the MACT bubble has been effective. Tests verified that the vent capture system is controlling substantially more HAPs (methanol) than would have been achieved by controlling all required sources at MACT standards. Plant engineers continue to seek other ways to further reduce HAP emissions. Pollutant discharges, such as biological oxygen demand effluents and total suspended solids, have remained below the new, more stringent limits set in the XL permit. Energy use has declined, and it is now very close to the target set in the FPA. Criteria pollutant emissions have been reduced and remain well below the various caps. The goal of a reduction in water use by one million gallons a day was finally achieved, and—as promised in the FPA—the XL permit limit for Flint River water usage was lowered by this amount in August 2000. Solid waste reductions were about halfway to the facility's goal of a 50% decrease from baseline, and studies are underway to find ways to reduce the amount of solid waste (lime mud) further.

However, Weyerhaeuser's feasibility studies on the project's ambitious long-term goal of cutting bleach plant effluent in half continue to be very disappointing. There were two problems. First, there was a limit to how often these waters could be recycled. Metals build up in the water with each recirculation, leading to the possibility that increasing concentrations of undesirable metals would appear in the fluff. Second, consumer demand for the super-white product had increased, thereby adding to the need for bleach plant flows. The approach examined by company engineers and scientists would have raised the cost of the product; when asked, Weyerhaeuser's customers said they were unwilling to pay a higher price for the fluff in exchange for this environmental benefit. These studies have resulted in unexpected spin-offs, however, namely, improvements in operations that should lower effluent discharges and reduce the amounts of lime mud produced (see Weyerhaeuser XL progress reports, www.epa.gov/projectxl/weyer).

Would EPA headquarters (particularly OPPE) have approved Weyerhaeuser's proposal without the goal of reducing bleach plant flow by 50%? EPA headquarters was hoping for closer to a 100% decrease, and thus the 50% target was from the beginning a compromise accepted by both sides. Certainly there was plenty of SEP contained in the rest of the project but, given its record, the company probably would have implemented most of this SEP under the conventional regulatory regime. The Weyerhaeuser FPA without the bleach plant flow reduction pledge was nevertheless a worthwhile experiment to see how added operating flexibility and compliance cost savings affected the environmental performance of such a facility.

As long as EPA was looking at Project XL as a quid pro quo—this much SEP for this much regulatory flexibility—worthwhile projects like Weyerhaeuser's faced obstacles in gaining EPA approval. EPA did approve the project, thanks in part to the strong support from Region 4 and the Office of Air Quality Planning and Standards, but also because it believed the 50% reduction in bleach plant flow was feasible. Whether it actually is feasible remains to be seen, however. Weyerhaeuser engineers, as a result of their search, however, have discovered process improvements that will yield tangible environmental benefits. This is an example of how nonenforceable "stretch" goals can lead to unexpected payoffs of SEP that more conservative guaranteed benefits might not have achieved.

Merck's Stonewall Plant, Elkton, Virginia

Merck's Stonewall facility was principally engaged in batch processing: fermentation, solvent extraction, organic chemical synthesis, and finishing operations employed in the production of pharmaceuticals. It was located four miles south of Elkton, Virginia, a small town of roughly 2,000 residents within Rockingham County, which has a population of about 68,000. The plant was established in 1941 and had about 800 employees in March 1997. Like Weyerhaeuser's plant, Stonewall was in an attainment area but within two kilometers of Shenandoah National Park. Air emissions of VOCs and criteria pollutants were the focus of the project, and emission caps were involved. Shenandoah National Park had been designated a visually impaired mandatory Class I area under the Clean Air Act as amended in 1997. The Clean Air Act amendment defines mandatory Class I Federal areas as national parks (over 5,000 acres) and international parks—all of which have been in existence as of August 7, 1977, and have been identified by the secretary of the Department of the Interior as having visibility as an important value (Clean Air Act, Sec. 162a). Section 169A(a)(1) of the Clean Air Act set goals for these areas: "Congress hereby declares as a national goal the prevention of any future, and the remedying of any existing impairment of visibility in mandatory Class I Federal areas which impairment results from manmade air pollution."

In 1990, the park management declared that visibility was seriously degraded, streams and watersheds were being acidified, and ozone and sulfur dioxide were injuring vegetation (March 31, 1997, *Federal Register* Notice, 62 FR 15304). Thus, like Intel, Merck was in a sensitive region where additional pollution had to be carefully monitored.

Substance

Manufacturing flexibility and rapid response to change were high priorities at Merck, as they were in the other cases we have examined. As Merck's original August 1995 XL proposal put it:

> Speed to market for new products, and new claims for existing products, is at the heart of the company's need to have flexible manufacturing facilities that can make a broad range of products in the same equipment, using a wide array of raw materials and solvents. Thus the ability of Merck's manufacturing plants to respond to rapidly changing market conditions and product demands is critical to Merck's ability to stay competitive in a worldwide pharmaceutical industry.

Combined with this need for flexibility, the company's corporate culture, according to its XL proposal, encouraged the search for continuous improvement that yields frequent manufacturing changes in existing products. The Stonewall plant in Virginia had "a Technical Operations Group of 75 scientists, engineers, and technicians" that set productivity and process improvement goals, which when achieved often reduced emissions per unit of product produced. As a result of these manufacturing changes, Stonewall and Merck's other facilities were likely to modify environmental permits frequently. To avoid a competitive disadvantage, Merck argued in its proposal that it was necessary for flexible manufacturing facilities to have flexible permits.

Environmental Regulatory Requirements. The plant, near a Class I area, was significantly affected by PSD requirements, which could call upon Stonewall staff to evaluate net emissions increases caused by physical or operational changes contemplated. These manufacturing changes had to be reviewed and approved by the Virginia Department of Environmental Quality. The state agency's review would be done in consultation with the National Park Service. Under PSD, a facility must compare *potential* emissions after the change with *actual* emissions before the change. The Stonewall plant could be discouraged from making process changes, including pollution prevention projects, if this actuals-to-potentials analysis would trigger PSD reviews.

Merck was concerned with the complexity of the air regulations that currently applied or would soon be applied to the facility. In addition to the PSD requirements, Stonewall was subject to Virginia BACT requirements, MACT standards for the industry and for industrial boilers, and Resource Conservation and Recovery Act Organic Air Emission Standards. The company summed up its concerns in its original XL proposal:

> Taken together, the interplay of the PSD regulations, the Title 5 operating permit rule, the MACT standards, and the HAP modification rule will create layers of regulatory requirements that are likely to reduce Stonewall's flexibility in the future. Therefore, Merck is interested in developing an alternative permitting strategy that produces greater environmental benefit than the sum of the existing and future potential regulations, while giving the plant the flexibility to respond quickly to market demands and to provide long term growth opportunities for the Stonewall plant.

The plant also needed a Clean Air Act Title 5 operating permit, and it would most likely be subject to a HAP modification rule.

The Primary Pollutants at Stonewall. The Stonewall facility was a source of criteria air pollutants: sulfur dioxide, nitrogen oxides (NO_x), carbon monoxide, and particulate matter. Before Project XL, the predominant source of these pollutants was the emissions-controlled coal-fired boilers at the plant that generated process steam. The criteria pollutant, ozone, was not emitted directly by the facility. However, the VOCs emitted by the pharmaceutical production units in the plant could play a role in the formation of ozone in the atmosphere and therefore were included as a criteria pollutant proxy for ozone. Stonewall also emitted some HAPs, both inorganic (hydrogen chloride and fluoride) and organic (chiefly methanol), but not in particularly large quantities (only about 7% of the total HAPs emitted at 3M's Hutchinson facility).

Because Shenandoah National Park was suffering degradation from ozone and acid deposition, the key pollutants of concern for Merck's XL project were sulfur dioxide and nitrogen oxides. Sulfur dioxide contributed to visibility problems, nitrogen oxides contributed to ozone formation, and both contributed to acid deposition in the park. However, the contributions to these problems from Merck's emissions were not believed to be among the more significant ones (Tedd Jett, Merck & Co., private communication, January

2001). In principle, the plant's VOC emissions could also have contributed to ozone formation, which is formed when VOCs and nitrogen oxides react in sunlight. However, scientific analyses indicated that the area's atmosphere already had sufficient concentrations of VOCs present to react with all the nitrogen oxides present, even without Stonewall's VOC emissions. The area was therefore designated to be "NO_x limited," meaning that ozone formation depended on additional NO_x emissions and not on additional VOC emissions. Gaining a consensus agreement on this point was a crucial element of the project.

Because production levels were below average for the years 1994 and 1995, annual air emissions averaged during the period 1992–1993 were established as the baseline from which to measure the environmental performance of the facility under an XL permit. Sulfur dioxide emissions, for this baseline, were about 720 tpy.[11] NO_x emissions were about 290 tpy, VOCs were 408 tpy, and PM_{10} and carbon dioxide emissions were 38 and 43 tpy, respectively. Total criteria pollutant air emissions (including VOCs) were slightly greater than 1,500 tpy.

Environmental Benefits of the Project. The Stonewall plant's coal-fired boilers were installed in 1982 with a useful life of 40 years. Nevertheless, Merck proposed to replace these boilers with low NO_x, natural gas–fired boilers at an estimated cost of $10 million. In addition, Merck had estimated increased fuel costs of $1 million a year (*Federal Register* 62: 15304, March 31, 1997). Merck argued that it could only afford to make this investment (and increase both its fuel costs and the uncertainty of future fuel costs) if it were granted substantial regulatory flexibility. The boiler replacement would yield substantial environmental benefits. Merck estimated that sulfur dioxide would be reduced 94% to 40 tpy, NO_x reduced 87% to 37 tpy, and PM_{10} reduced 88% to about 5 tpy. Also, because burning coal was the major source of the plant's inorganic HAP emissions, HAPs would be lowered 65% to 25 tpy. VOC emissions, coming from production units rather than the boilers, would remain unchanged.

Air Emissions Caps and Regulatory Flexibility. Merck originally proposed a single cap on the total air emissions of VOCs, sulfur dioxide, nitrogen oxides, carbon monoxide, and PM_{10} combined and agreed to replace the coal-fired boilers with natural gas units and/or pursue other environmentally beneficial projects. The company's objective was to be able to trade the reductions in sulfur dioxide and nitrogen oxides realized by the conversion to gas for increases in VOCs—something that could be expected as production at the plant increased over time. After negotiations with the stakeholder group, Merck agreed to a more complex series of caps that would take force shortly after the boiler replacement had been completed (Table 8-5).

Table 8-5. Merck's Stonewall Plant Criteria Pollutant Emissions (tons per year)

Pollutant	Baseline (1992–93 average)	Post-boiler replacement (estimate)	Plantwide pollutant emissions limits (caps)	Percentage of baseline
Sulfur dioxide	719	40	539	75
Nitrogen oxides	291	37	262	90
Particulate matter 10 microns or less in diameter	42	5	42	100
Volatile organic compounds	408	408	None	
Carbon monoxide	43	50 ± 10	None	
All of the above	1,503	540 ± 10	1,202	80

Source: Merck XL Permit and Final Project Agreement.

The XL permit allowed "Merck to construct or modify emission units at the site" without the need to satisfy the various preconstruction permitting requirements for the capped pollutants. In addition, the Stonewall facility could make "significant" modifications or new installations of a "process unit" without permit reviews, provided it installed "at the process unit emission controls, pollution prevention or other technology that represents good environmental engineering practice" for the industry. In the event an applicable and new criteria air pollutant regulation was promulgated by the EPA or Virginia, Merck could choose to either satisfy the new requirement or calculate the amount of emissions reductions that compliance with the new regulation would yield and, instead of complying, reduce the appropriate cap (or caps) by that amount. Merck had to submit this calculation for review to the agency administrating the new regulation (EPA or Virginia Department of Environmental Quality). Merck argued for this provision to cover cases when it could achieve equivalent emissions reductions elsewhere at the plant over time more cost-effectively than complying directly with the regulation or when a particular affected production unit was being phased out in the foreseeable future.

As Table 8-5 shows, replacing Stonewall's coal-fired boilers with natural gas–fired boilers with low-NO_x burners would reduce sulfur dioxide, nitrogen oxides, and particulate matter emissions to levels far below their individual caps. As a result, the sum of the criteria air pollutants and VOCs would be roughly 650 tpy below the cap of 1,202 tpy. Theoretically, this large gap between post-replacement actual emissions and the cap for total pollutants would allow the facility to increase VOC emissions by 650 tpy as a result of expansion and increased production of existing products. EPA and the Virginia Department of Environmental Quality were trading increased emissions of VOCs (caused, potentially, by product mix changes and new product introductions at the plant) for substantially reduced emissions of the two criteria pollutants, sulfur dioxide and nitrogen oxides, that were believed to be most responsible for the degradation experienced at adjacent Shenandoah National Park.

Superior Environmental Performance. Merck began to develop its XL project with two key elements well established for its Stonewall Plant: a well-defined emissions baseline from which to measure environmental performance and a much greater environmental benefit from sulfur dioxide and nitrogen oxide reductions than from VOC reductions. The EPA supported its decision to approve the Merck XL project by arguing that the area where Stonewall was located had "been well documented to be NO_x limited for ozone formation" (*Federal Register* 62: 52629, October 8, 1997). Consequently, the EPA strategy was to favor a reduction in NO_x emissions over reductions in VOCs, with the additional benefit that NO_x also contributed to acid deposition in the park.

EPA also argued that Merck had strong incentives to limit increases in VOC emissions as well. For example, the XL permit required Stonewall's compliance burden to increase as total criteria and VOC emissions approached the cap. Furthermore, the large gap between estimated actual emissions and the cap provided Merck with a degree of flexibility the company was not likely to squander with large increases in VOC releases to the environment. The increased accessibility to pollutant emissions data that the public would have and the permit review by the stakeholder group required every five years also provided incentives for Merck to minimize increases in VOC emissions as much as possible. EPA also pointed to a worst-case National Ambient Air Quality Standard analysis of the impact of Stonewall's increased VOC emissions on ozone formation and found it to be small and, in other models, negligible (*Federal Register* 62: 52627, October 8, 1997). Finally, Stonewall's XL permit required Merck to adopt measures that followed good engineering practices to control VOC emissions when a process unit modification or new process unit increased the potential of the facility to emit VOCs by 40 tpy.

EPA noted that because Stonewall was in an attainment area, under conventional regulations it could increase emissions by expanding production on existing facilities or, after obtaining the proper permits, by adding new equipment. In exchange for no longer obtaining preapproval for changes at the plant, Merck had agreed to a sharp reduction in its allowable criteria pollutant emissions. For example, the allowable level of total criteria pollutants (and VOCs) emissions went from 2,700 tpy to the cap of 1,202 tpy. The allowable on NO_x went from 569 tpy to the cap of 262 tpy (*Federal Register* 62: 52626–27, October 8, 1997). EPA differentiated the caps in the XL permit from a plantwide applicability limit (PAL) (*Federal Register* 62: 15304, March 31, 1997). Under a PAL, a plant could make modifications that would take its emissions over the PAL as long as the facility could successfully complete a PSD application or new source review to obtain the necessary permits. The caps for Stonewall could not be violated under any circumstances. In summary, EPA argued that Merck's proposed FPA offered a significant, highly predictable level of SEP. There was no question that the change from coal to natural gas would result in large reductions in sulfur dioxide, nitrogen oxide, and PM_{10} emissions. VOC emissions could increase but were not likely to do so by significant amounts. As long as the area remained NO_x limited, this trade-off of some potential increases in VOC emissions for very large decreases in sulfur dioxide, nitrogen oxide, and PM_{10} emissions provided a clear environmental benefit.

Process

Merck began reaching out to stakeholders early in the process when its initial proposal was taking shape in mid-1995. The company shared its ideas for the project with OPPE at EPA headquarters, the Virginia Department of Environmental Quality, and Merck's Community Advisory Panel, and it received encouragement. In November 1995, Merck's proposal was among the first eight XL projects chosen, along with those of the Minnesota Pollution Control Agency, Intel, 3M, and others. Merck started work on the project in January 1996, and, at a large group meeting in February 1996, formed a representative stakeholder working group. In addition to Merck, the stakeholder group included representatives from the Virginia Department of Environmental Quality, the EPA Region 3 office in Philadelphia (one from Air Permitting and another from the Office of Regional Counsel), EPA's Office of Air Quality Planning and Standards (OAQPS), the Air Quality Division of the National Park Service in Washington, D.C., the air quality specialist from Shenandoah National Park, and three local community representatives. The local stakeholders were two residents living near (downwind most of the time) Stonewall and a member of the Rockingham County Board of Supervisors.

The group met, with few exceptions, every week, alternating one week in person and the next week by conference call. For the face-to-face meetings, the two EPA Region 3 people drove the 200 miles or so to Virginia from Philadelphia, and the OAQPS representative drove a similar distance from North Carolina. This participation showed a strong commitment to this project by Region 3 and OAQPS. EPA headquarters staff did not participate directly in the stakeholder group's deliberations, but issues were raised that the EPA representatives took back to Washington. According to a Merck participant, Region 3 was able to get a fast response from EPA headquarters, and the presence of Region 3 and OAQPS at the negotiating table helped greatly in reaching agreement (Tedd Jett, Merck & Company, private communication, September 2000).

Although not direct participants in the design process, several environmental advocacy organizations—including the Southern Environmental Law Center, the Virginia Consortium for Clean Air, and, to a lesser extent, the NRDC—had input during the project design. Perhaps most active was the Southern Environmental Law Center, which had a membership of more than 2,000. David Carr, staff attorney with the center, worked with the two local residents in the stakeholder group and provided them with technical assistance. The residents complained that they were not experts on the subjects in question and needed independent expert assistance to fairly evaluate the difficult issues that arose. The Rockingham County Board of Supervisors' representative found himself in much the same situation, and the board hired a consultant from nearby James Madison University (transcript, public hearing by Region 3, Harrisonburg, Virginia, April 14, 1997).

When the project had begun to take shape, OPPE organized a meeting on May 23, 1996, to brief national and regional environmental advocacy organizations, social and environmental justice organizations, people at OPPE, and others about the Merck project. The meeting was similar to the ones EPA organized for Intel and Weyerhaeuser. All of the meetings were facilitated by the Keystone Center, a Colorado organization with experience in conducting meetings concerning the environment. EPA did not organize such a meeting for 3M. Among the issues raised by the NGOs at the Merck meeting were the fact that the local stakeholders would not be signatories to the FPA and therefore had insufficient power; the nature of current health risks and the possible shifting, under the cap, from one VOC pollutant to a more harmful one; the indefinite term of the XL permit and FPA; and the possibility that Merck might be exempt from future regulations that, if applied to the facility, would require a lowering of the caps.

The Negotiations. According to Tedd Jett, Stonewall's environmental manager and participant on the stakeholder group, nearly all provisions of the project had been agreed to by the group by September 4, 1996, and the remaining issues were resolved by December 12, 1996. Included in the issues resolved by the stakeholder group were the indefinite length of the permit, the levels of the caps (the NO_x cap and the PM_{10} cap presented the greatest difficulty), and acceptance by all parties, including the National Park Service, that the area was NO_x-limited for ozone formation due to the amount of VOCs generated by biogenic sources (such as coniferous trees). Compromises clearly had been made on all sides. Merck had wanted only a single cap for total emissions but was convinced by other stakeholders that individual pollutant caps were needed to ensure SEP. Merck agreed to extensive reviews and analyses, which included periodic reevaluations of the NO_x-limited status of Shenandoah National Park; to modeling non-HAP VOC concentrations near the plant to ensure that Virginia's air toxic standards were not exceeded; and to measurement, recording, and reporting requirements that, according to the EPA Region 3 administrator, were more stringent than Merck's previous permits (Tedd Jett, Merck & Company, private communication, September 2000; letter from Tedd Jett to EPA Region 3, May 14, 1997).

Merck, conversely, was able to get a fair amount of flexibility under the facility's air emissions caps, including being able to rely on its engineers' own judgements as to good engineering practice instead of undergoing the usual BACT regulatory analysis. The NO_x and PM_{10} caps were set well above the emissions levels projected after boiler replacement. In both cases, there were reasons for doing this (for example, PM_{10} emissions concentrations were so low the actual numbers were uncertain), and the compromise levels were worked out in the process of give and take that took place during the many stakeholder group meetings.

Although a consensus agreement had been reached, the local citizen stakeholders were never completely happy with the role they would play in the group's deliberations during the five-year reviews required by the permit. The local representatives were not signatories of the FPA, and though their views would receive due consideration during the reviews, only the five signatories (representatives of Merck, EPA Region 3, the National Park Service, Virginia Department of Environmental Quality, and the Rockingham County Board of Supervisors) had to agree to permit changes suggested by the review. The neighborhood representatives wanted one vote, arrived at by consensus among themselves.

Another issue that needed resolving concerned SEP. Was it superior enough? According to a Merck participant, EPA did not question the level of SEP to any extent. Park representatives and the local stakeholders raised this issue most often (Tedd Jett, Merck & Company, private communication, September 2000). EPA was comfortable with the large near-term reductions in sulfur dioxide and nitrogen oxide emissions and with their long-term permanent reductions. The agency was in a position to defend the project against the objections raised by NGOs and others—and it did so.

All did not go smoothly within EPA, however. Site-specific rulemaking was a contentious process, and although the stakeholder group had reached agreement by mid-December 1996, the first *Federal Register* Notice proposing the permit language was not published until March 31, 1997, and a second, final Notice not until October 8, 1997. The actual permit issued by the state became effective February 10, 1998. The Merck participants were worn out and wondered during the second half of 1997 whether the permit would ever be approved. A major part of the second year was devoted to responding to a large number of comments and to the extra care taken by both sides to construct a foundation for the permit that could stand up to any legal challenge (Tedd Jett, Merck & Company, private communication, September 2000).

The inclusion of representatives from EPA Region 3 on the stakeholder group was a crucial element in reaching a final agreement, and the participation by OAQPS was helpful as well. For example, NRDC submitted detailed comments on the Merck project, as well as on the Intel and 3M projects, in a July 3, 1996, letter to EPA. The Region 3 administrator, in an August 15, 1996, letter to NRDC in defense of the project, gave an extensive response to the objections and questions raised by NRDC. Clearly, EPA wanted an agreement in the Merck case, and it did what it had to do to assure that an agreement was forthcoming.[12]

The Experimental Nature of the Projects

Although these three cases, for which agreements were reached, had a number of similarities to the 3M case, they had notable differences as well. A very important difference, in our view, was that they were not as experimental as the 3M project. Intel, for instance, probably could have achieved the same type of agreement without XL. Though Ocotillo was in a nonattainment area, the flexibility the company obtained had little to do with the environmental performance to which it was committed. Unlike the other cases, Intel did not need a site-specific XL permit. Total emissions at its fabrication plants were to be less than required for a single facility to be designated a minor source. Consequently, it was likely that Intel could have negotiated with Maricopa County Department of Environmental Quality and EPA Region 9 for a permit comparable to the one it received under XL.

Similarly, Weyerhaeuser's XL permit, though broader and more comprehensive than Intel's, was not particularly experimental. A program to achieve reductions in pollutant discharges into the Flint River already was in place prior to approval of Weyerhaeuser's XL permit and would have continued and functioned at a similar level even if Weyerhaeuser had not obtained an XL permit. The company had many improvements in the pipeline that were destined to yield additional reductions in pollution even without its application for a permit under XL. The project would test the workability and economic benefits of plantwide air emissions caps, but it would be difficult to determine if this facilitywide, performance-based approach would stimulate environmental performance superior to what conventional command-and-control regulation would have achieved. There were some additional incremental gains to the environment guaranteed by Weyerhaeuser's XL permit, but the substantial benefit offered by the 50% reduction in bleach plant flows was an uncertain stretch goal that at this writing seems unlikely to be achieved.

A similar argument can be made about the Merck facility. Merck's permit did not induce significant technical or management innovations beyond what would have taken place outside the XL framework.[13] Boiler conversions to low-NO_x natural gas–fired fuel that resulted in large reductions in criteria pollutants were common. They were carried out at many companies without the inducement of XL. Absent new regulations, however, it is fair to say that given the cost, Merck would not have made the change without the permit.[14] 3M had replaced its coal-fired boilers with natural gas–fired boilers many years earlier.

In contrast to the proposed 3M XL pilot, we would argue that the Intel, Weyerhaeuser, and Merck projects were less experimental. At the same time, there was less guaranteed SEP in the proposed 3M agreement. One of the aims was to measure the connection between flexibility and SEP.[15] The FPA stipulated that 3M had to estimate the percentage of a project's results that could be attributed to the permit's flexibility. These differences and how they yielded different results will be explored further in the next chapter.

Notes

[1]Information about these pilots can be obtained from documents available at EPA's Project XL website. These include the signed FPAs, FPA applications, and *Federal Register* Notices for these projects (no *Federal Register* Notice was required for the Intel project); comments and letters pertaining to each project; progress reports; and so on. We have also benefited from several studies of XL projects, the most notable being the National Academy of Public Administration Report (NAPA 1997). Finally, we have interviewed each facility's environmental manager and, in Intel's case, the stakeholder representative from the Maricopa County Environmental Services Department, the air emissions regulatory authority for facilities in the Phoenix area.

[2]According to Delmas and Mazurek (2001), the Intel project took 17 months, the Weyerhaeuser project took 16 months, and the Merck project took 19 months. We say that these projects took well over 11 months to negotiate because starting points are hard to determine with certainty and there were hiatuses in the middle of the negotiations when the federal government was shut down because of budgetary issues.

[3]Here the term *source* refers to an entire facility producing a particular type of product. See Table 8-2 for federal minor plantwide source requirements.

[4]Under a conventional Maricopa County Environmental Services Department air permit, Intel engineers needed preconstruction approval before installing new equipment. Given the rapid change in semiconductor production technology, older and usually nonequivalent equipment had to be used to estimate pollutant emissions expected from the new equipment. Maricopa permit engineers would review the application, including the planned emissions controls. After gaining the county's approval, Intel would buy and install the new equipment and develop an accurate assessment of air pollutant emissions, which it would then submit to the county for re-review. Important changes in existing equipment or in the process (such as chemical changes) had to go through a similar double review process. Maricopa County's conventional regulations for the industry did have some flexibility. The many small changes that Intel would make during the year needed only to be recorded in a log, which the county might periodically review.

[5]At the time this book was written, Intel had completed a low-cost expansion of the manufacturing capacity and floor area at Fab 12 and began operating a second, more advanced facility, Fab 22, in 2001. The five-year FPA was renewed for another five years early in 2002. The new FPA noted that additional semiconductor-related facilities may be built on the site and, to accommodate this, the cap for VOCs would be increased from 40 to 49 tons per year. Caps for sulfuric acid and phosphine were both lowered to only one ton per year in the new agreement.

[6]*Production unit factor*, PUF, was defined using appropriate units as the area of silicon processed divided by the line width of the smallest transistor on the chip (for example, 0.35 microns in 1997, 0.13 microns at the new Fab 22).

[7]For a more detailed account of the "end run," see NAPA 1997, 92–94.

[8]According to Intel's December 1999 XL Progress Report, the flexibility granted it in its XL permit eliminated 30 to 50 permit reviews a year. The reporting format initiated by the stakeholder team for the project (one annual and four quarterly consolidated reports) was recommended by EPA for adoption by others in the agency's

"Guide to XL Project Teams—Project Tracking and Reporting." Emissions in 1998 from Fab 12 of sulfur dioxide and PM_{10} were far below the low limits set by the permit, and VOCs, carbon monoxide, and nitrogen oxide emissions were well below their caps. Although HAP emissions increased in 1998 compared with 1997, production increased even more. Consequently, total HAP emissions per unit of production decreased by 30%. During the five-year period, 1997–2001, emissions of VOCs, HAPs, and criteria pollutants remained consistently well below their respective limits. If measured per unit of production, comparing 2001 with 1997, HAPs per PUF decreased by 45% and VOCs per PUF by 78%. No detectable amounts of phosphine and sulfuric acid were emitted. A 14-pound-per-year cap for arsine was added in 1998; 2.8 pounds of the chemical was emitted in 2001. Both solid waste and hazardous waste recycling goals were being met, and water conservation and reuse was going well (Intel Project XL, December 1999 and 2001 progress reports).

[9]An air quality analysis showed that when emissions remained below the level of the caps, there would be no violation of National Ambient Air Quality Standards or other requirements (Appendix 6, Final Project Agreement; *The Weyerhaeuser Company: Frequently Asked Questions about the XL Project*, www.epa.gov/projectxl/weyer/011797_f.htm, accessed June 18, 2002).

[10]The predominant HAP emitted at the Flint River Plant was methanol, which was considered to be less toxic than the predominant HAPs emitted by the 3M–Hutchinson facility.

[11]To put this number in perspective, a 500-megawatt electrical energy generating plant that consumed very low sulfur coal (0.5 lb. of sulfur per million Btu of coal), easily satisfying the more stringent requirements of Phase II of the Clean Air Act, would emit 720 tons of sulfur dioxide, running full out, in less than 13 days.

[12]As a final note, after some technical delays, the boiler replacement was completed and up and running by the first week of July 2000. By November 2000, criteria pollutant emissions for the year had dropped below the XL caps, well ahead of the July 2001 deadline.

[13]Under its XL agreement, Merck actually could increase its emissions after the boiler conversion took place. Nonetheless, in this case as in the others, there was the possibility that some learning might take place under XL that otherwise might not take place. In the Merck and Weyerhaeuser cases, this learning would be related to how the companies' engineers would respond to the plantwide standards and to the increased operating flexibility.

[14]The same result, perhaps, could have been achieved by simply tightening regulations in light of the national park's worsening air quality and increased acid deposition (the growth in nitrogen oxide and sulfur dioxide emissions). However, had the Virginia Department of Environmental Quality required these reductions, Merck probably would have made sure only to make high-value-added products at Stonewall, or it might have moved production elsewhere out of a Class I area into an attainment area, possibly closing the plant entirely.

[15]Some mandated SEP existed. 3M agreed to BACT in cases of sources with greater than 100 tpy of VOC emissions, and additional SEP would be required if during the later years of the permit production increases put pressure on the relatively high caps.

9

Comparing the Approved Projects with 3M's Proposal

In this chapter, we compare the proposed 3M–Hutchinson Project XL pilot with the Intel Corporation, Merck & Company, and Weyerhaeuser Company pilots to explain why 3M Company failed to reach an agreement when the other firms did. Two key factors differentiate Intel's, Merck's, and Weyerhaeuser's projects from 3M's proposed pilot: first, the near certain superior environmental performance (SEP) offered in the Intel, Merck, and Weyerhaeuser cases in contrast to the uncertainties in the 3M project and, second, the negotiating process itself, most notably the key role played by the U.S. Environmental Protection Agency (EPA) regional office in the Intel, Merck, and Weyerhaeuser cases and the broad coalition in which the regional offices participated, in contrast to the main role played by the Minnesota Pollution Control Agency (MPCA) in the 3M case and the narrower coalition that MPCA assembled. Because EPA had approved MPCA's proposal to be an XL participant, the state agency thought it might be able to make decisions on its own. Consequently, 3M did not deal directly with EPA, the rule setter, but with state government, and state government was fairly powerless in determining the rules by which the game was played and in shaping the results.

Why an Agreement Was Not Reached in the 3M Case

Collaborative endeavors result in an agreement or lack of one for many reasons. Unpredictable and random events may play a role. These cases are no exceptions to this rule. In the three cases where an agreement was reached, the outcome was by no means a foregone conclusion. Serious roadblocks materialized after the stakeholders had worked out an agreement, and there was a time in each case when the company participants thought about walking out.[1] Moreover, it is conceivable that 3M, had it the time and inclination to do so, could have reached agreement with the EPA to implement a scaled-back, less ambitious project. Instead, 3M ran out of the time it was willing to commit, and it found the alternative that EPA was proposing to be risky, cumbersome, hard to carry out, and not very interesting. Thus, 3M followed through on its intent to withdraw, whereas the other companies that considered this possibility did not walk away from the negotiations. Still, this possibility aside, *in all four cases it was much more difficult to achieve a successful outcome* than could have been expected after stakeholder groups reached an initial agreement.

The significance of delay in these cases and others is that implementation of Project XL on a broad scale is probably unfeasible (Delmas and Mazurek 2001). XL is restricted to firms that find the costs low enough or the benefits high enough to justify making the considerable effort required. A sample this small is not sufficient to experiment with interesting policy options (Blackman and Mazurek 1999). Controlling costs and minimizing transaction times, therefore, are key challenges that XL or programs like it must confront if they are to be viable in the future. Firms that are aware of the transaction costs of XL will be eager to participate only if these issues are addressed.

The Projects Examined in This Book

The four projects examined in this book have similar elements. All of the firms had corporate policies of continuous improvement in production and environmental performance. All, to varying degrees, needed rapid responses to market, product, and technology changes. Intel, Merck, and 3M wanted regulatory flexibility that would better accommodate growth in production levels at the project facilities. All of the firms maintained that air pollutant emissions regulations posed barriers, and consequently they sought regulatory relief from the need for frequent air permit reviews. In each instance, the approach adopted was to use facilitywide or sitewide air emissions limits to provide this flexibility, and these limits generally were set largely well below what current regulations allowed.

There were important differences, however, between the 3M proposal and the other projects. Unlike the other projects, 3M had many emissions units that were grandfathered in and therefore unregulated, a situation not uncommon in industry. Nevertheless, the company had voluntarily applied emissions controls and recycling technology to most of these grandfathered units, thereby creating an enormous gap between the total emissions of volatile organic compounds (VOCs) allowed by the regulations and the facility's actual VOC emissions. In addition, production at Hutchinson had been increasing steadily, and 3M was planning for this rate of increase to continue over the 10-year life of the XL permit it was seeking. Consequently, to allow for this increase in production, although the VOC cap was set considerably below current regulatory requirements, its level was almost double that of current VOC air emissions, which caused problems for EPA. On top of this, the high degree of uncertainty over Hutchinson's production plans was unsettling to the agency. Overall, the uncertainty of 3M's proposal raised issues for which EPA was unprepared.

Production increases were an issue in several of the other projects, but the problem was easier to manage. Intel planned and carried out major production increases at its Ocotillo Campus, but the company had been able to reduce its air emissions per unit of production to such a low level that its sitewide XL permit caps could accommodate this large expansion while maintaining a minor-source designation for the entire site. Output increases at Weyerhaeuser's Flint River plant were expected to be modest.[2] Merck's Stonewall plant also wanted to have air emissions caps that could accommodate production increases, which could have posed a greater problem than it actually did. However, because Stonewall's criteria pollutant caps were set well below baseline levels experienced in previous years, the facility did not face the objections 3M did.[3]

In addition to these differences in substance, the cases exhibited process differences. As has been noted, Intel, Merck, and Weyerhaeuser formed broader coalitions than 3M. They had more inclusive stakeholder processes. The focus for negotiating an agreement in the Intel, Merck, and Weyerhaeuser cases was the EPA regional office, whereas the focus in the 3M case was the state pollution control agency, MPCA. The 3M project was carried out by the state, whereas the EPA regional office, which proved to be a critical player in the other cases, was brought in very late. Similarly, a representative from the EPA Office of Air Quality Planning and Standards was an influential participating stakeholder in the Merck and Weyerhaeuser projects but not in the 3M case. EPA Region 5's culture, which was different from that of the other participating EPA regional offices, also may have influenced EPA's failure to reach an agreement with 3M. In any event, the coalition in favor of an agreement in 3M's case was more limited and in the end weaker.[4] These differences in substance and process were critical in the cases we examined, and they go a long way toward explaining the different outcomes.

Substantive Differences: Assured Superior Environmental Performance

Defining a Baseline

The Intel, Merck, and Weyerhaeuser XL projects had enforceable environmental performance requirements in their facilities' permits that ensured a measurable amount of SEP would be achieved. In each case, there was a well-defined baseline. Intel intended to operate its Ocotillo plants such that air emissions would remain below the threshold for triggering major-source (facilitywide) permit requirements.[5] These sitewide limits provided a baseline for the project. Weyerhaeuser's Flint River plant guaranteed SEP for effluent discharges into the river.[6] Given the plant's relatively stable production outlook and well-defined products, recent operations provided clear performance baselines for both air and water pollution. Merck's Stonewall plant also had reasonably well-defined baselines for its air pollutant emissions. Though Merck produced a varying variety of pharmaceutical products, their general characteristics were fairly well understood so that the output increases planned for the facility did not confront stakeholders with the kind of uncertainties that were present in the 3M project. Criteria pollutant emissions were due predominantly to the coal-fired boilers and were fairly independent of the production mix. Consequently, actual pollutant releases during a recent, typical operating year provided stakeholders with an acceptable baseline from which to measure SEP.

   3M's Hutchinson plant also had well-defined air emissions levels from its recent operating history. However, its VOC and hazardous air pollutant emissions baselines were not that relevant to future operations, given the large changes about to take place and the likelihood of substantial production increases involving a range of unpredictable products. Because Hutchinson had so many grandfathered emissions sources, its permit limits were so far above actual emissions as to be of little use in establishing a baseline. Meaningful definitions for emissions per unit of production were complex, and they were difficult to construct given the wide variety of products that would be produced, many of them new. What should the unit of production be? Simple metrics such as weight, length, or area of product could not easily accommodate the variety of shapes, sizes, and properties of the products produced there. Using pollution per unit of value added in dollars might be an interesting approach, although it would expose a company to price fluctuations caused by market forces. If 3M, MPCA, the Pilot Project Committee (PPC), and EPA had worked together, they might have come up with a suitable definition. But this approach was never seriously explored, primarily because 3M managers were looking for simplicity and were not interested in imposing greater complexity upon its Hutchinson engineering staff.

The SEP Offered

Each of the other cases offered clear SEP. Intel's project provided clear air quality benefits. Environmental regulations set minor-source emissions limits for an entire facility (e.g., no more than 50 tons per year of VOCs). Intel agreed to emissions caps for the air pollutants it produced at a level that was below these minor-source limits for total emissions *from all the facilities built at the Ocotillo site combined.* Because the company planned to build over time at least two large semiconductor fabricating plants, Intel's XL permit limits would restrict total emissions to be 50% or more below what the company needed to do to remain a minor source. Intel responded to other community concerns as well—for example, by protecting water quality and minimizing impact on the neighborhood.

Weyerhaeuser's Flint River project was more complicated but nevertheless also offered clear environmental benefits. The fluff-producing facility already was operating with an environmental performance record below regulatory requirements and was setting a standard for the industry. Nevertheless, the plant made the enforceable commitment to reduce effluent discharges per unit of fluff produced to below the already low levels of current operations. In addition, because production at the plant was not expected to increase very much over the life of the permit, absolute pollutant discharges also were limited. As part of the project's quid pro quo, Weyerhaeuser offered additional environmental benefits, such as reduced water use, improvements in managing its extensive Georgia timberlands, and a reduction in solid waste. The Flint River project also employed facilitywide air pollutant emissions caps (actually two sets of caps) that were set at 40% of the facility's previous permit limits.

The air emissions limits were somewhat similar to those in the 3M case, and one could argue, as 3M did, that these limits also constituted SEP. As in the 3M case, the caps were set higher than Flint River's actual emissions (roughly equal to existing VOC emissions and 43% higher than baseline sulfur dioxide and nitrogen oxide emissions). Altogether, the SEP contained in the Weyerhaeuser project's preliminary agreement was considerable, but nevertheless it was not strong enough for EPA headquarters. Weyerhaeuser had to offer more to clinch the deal. The company made the commitment to *try* to achieve a 50% drop in bleach plant effluent flow during a 10-year period.

By switching boilers from coal-fired to low nitrogen oxide, natural gas–fired, Merck's XL project provided clear environmental benefits by substantially reducing criteria pollutants. To offset the cost of conversion and the higher fuel costs with which Merck would have to cope, the company wanted to be able to boost production and therefore incur higher VOC emissions without costly permitting delays. Though this desire by Merck caused some controversy, the trade-off was determined to be a net environmental benefit for the region.

The 3M–Hutchinson proposed XL permit and final project agreement (FPA) did offer some environmental benefits. But in contrast to the three other cases, these benefits were more theoretical. Though most, if not all, of Hutchinson's unregulated production lines eventually would come under regulation, the XL permit in effect regulated these sources up front. As a result, the XL permit capped air emissions at levels far below what current regulations allowed. Unfortunately, the caps also were set well above current actual emissions and did not by themselves present a compelling enough case for guaranteed SEP.

Clearly, 3M did not offer and probably would have been unable to offer the kinds of predictable improvements in environmental performance that were present in all three of the other projects and that served as currency in the negotiations to obtain regulatory flexibility for SEP. 3M was not constructing a new plant that benefited—the way Intel's did—from improvements in pollution-preventing process technology and in control technologies. 3M did not make any offer concerning water conservation—as Intel did—because it did not need to, given that the region where it was operating did not have significant water problems. 3M was planning to add several new products and corresponding production lines as it gradually phased out magnetic tape manufacturing, which only added to the uncertainty about the facility's future environmental performance. It was going to introduce new pollution prevention technology in some of the adhesive tape producing units replacing the magnetic tape units at the North Plant. The company could have used the innovations it was likely to make as an SEP bargaining chip, but its managers' conception of an XL project did not include this type of a quid pro quo guarantee.

These important differences in the substance of the 3M project were major reasons for the failure to reach an agreement. The range of different tape-related products 3M produced did not lend itself to the introduction of a major process change that could yield a sizable reduction in emissions (in comparison with Weyerhaeuser's promised 50% reduction in bleach plant effluent flow). The magnetic tape facility's solvent-recovery system could have made such an impact, but 3M had already installed it some years earlier. 3M's Hutchinson plant already burned natural gas for its boilers, and thus it could not offer to convert to natural gas, as Merck did.[7]

In fact, 3M and, for the most part, the other Minnesota stakeholders were not interested in making a deal that 3M would provide a level of SEP for a level of flexibility. 3M and the other Minnesota participants had a different conception of Project XL. Because they were assured that Hutchinson's environmental performance would always be better than regulations required, their focus was on providing *flexibility as an incentive and facilitator of environmental improvements, including increased pollution prevention.* They regarded the 3M project as an *experiment* that, at little risk to the environment, would test several new and promising approaches to environmental regulation.[8]

Had the Minnesota stakeholders been able to arrive at sensible emissions limits per unit of production, a more acceptable standard of environmental performance for the project might have emerged.[9] In contrast, Weyerhaeuser could easily define its key pollutant releases per unit of production—for example, biological oxygen demand per ton of fluff produced. Similarly, Intel's facility produced Pentium microprocessors, whose rapid rate of technological improvement could be easily accommodated in the definition of a unit of production. Merck did not introduce the concept of pollution per unit of production, but its boiler conversion reduced emissions so dramatically that this measure was not needed, although it might have helped. There was some controversy among the stakeholders about the potential for large increases in Merck's VOC emissions. Setting goals in the FPA for decreasing VOCs emitted per unit of production might have made it easier to reach an agreement.

The Minnesota stakeholders thought that the project they had approved had a reasonable chance of providing SEP, but EPA wanted greater certainty that SEP actually would be achieved. By agreeing to the XL permit developed by the Minnesota stakeholders, 3M had basically accepted regulation of all of its grandfathered emissions units. Consequently, in practical terms, 3M could have accepted lower caps [for example, a VOC cap set at 3,500 rather than 4,500 tons per year (tpy)]. This limit would accommodate the expected production increases, provided there were some improvements in pollution emitted per unit of production. However, because this 3,500-tpy limit was well over the current emissions level of about 2,300 tpy of VOCs, EPA would probably have remained unsatisfied.

This approach was never seriously explored. Given the large amount of uncertainty about the many changes about to take place at the Hutchinson complex, 3M was reluctant to give up much more of the cushion between its actual emissions and allowed emissions than it had already agreed to. Because this cushion was created in large part by 3M's voluntary initiatives, the gap between the caps and actual emissions, in EPA's view, constituted "credit for past good deeds" and provided a license for 3M to backslide. In interviews with EPA staff, the view expressed was that 3M was going to use the cap to sharply increase emissions at the plant. EPA, MPCA, and, somewhat less enthusiastically, 3M spent the last few months of 1996 trying to negotiate a way out of this impasse but ultimately were unsuccessful.

The Process Differences: A More Inclusive Negotiating Process

Were the disparities between EPA's vision and goals for XL and those of the Minnesota stakeholders too great to have been bridged? Could a different stakeholder and negotiating approach have produced a different outcome? Of course, we will never know. We feel, however, that a different process would have increased the odds of achieving at least some sort of agreement, albeit one much less ambitious in scope.

The three stakeholder processes in the comparison cases we have reviewed differed in important ways from the stakeholder process in the 3M case. The approach they adopted was to create a stakeholder group that included the EPA regional office along with other government and citizen representation, among which were all the parties that had to officially agree to and sign the FPA. This meant that the state pollution control agency as well as the appropriate EPA regional office had to be represented on the team. City and county government offices were also represented. The EPA regional offices committed experts, who actively participated on the stakeholder teams from the beginning. This was not easy, because in each case the regional office was not close to the project facility. Nevertheless, EPA regional office staff members made the effort to attend many of the meetings. Regional office resources—funds for travel and staff time—were committed to these XL projects, most particularly in the Intel and Merck cases.[10]

Having staff from the regional office participate rather than staff from EPA headquarters was an important choice.[11] It made sense because regional staff had better information and better access to information than headquarters. The regional office staff worked closely with the environmental protection departments and industrial facilities in the states under their jurisdictions, whereas the main focuses of headquarters were on national policy issues and on setting standards. Each XL pilot was site specific and involved highly technical issues and local concerns. But, each project encountered problems when EPA headquarters staff entered the picture and raised the stakes by introducing national policy and political concerns.

The stakeholder members representing the EPA regional offices—and for the Merck and Weyerhaeuser pilots, the Office of Air Quality Planning and Standards—participated in the give-and-take required to reach a consensus on the design of an FPA. At the same time, they established effective communication with their superiors at their offices and at EPA headquarters. They understood the rationale behind the many compromises and concessions the stakeholders made. After a final consensus had been reached, armed with predictable and worthwhile direct environmental benefits to support their case, they were able to effectively defend that consensus against opposition both within and outside EPA.

In 3M's case, the stakeholder process was different in a number of ways. Most obvious and most harmful to the efforts to achieve a consensus was the absence of participation by EPA Region 5 staff in the project design phase during the first half of 1996. The problems that resulted from Region 5's absence became apparent during the late May 1996 meeting in Chicago of the Minnesota stakeholders with Region 5. Officials and staff from the agency had not been part of the give-and-take that led to the proposed FPA, nor had they been able to properly vet with headquarters any of the issues that arose. Contrast this situation with the fact that one and two weeks earlier in May, Intel, Merck, and Weyerhaeuser had presented the outlines of their projects to nongovernmental organizations and EPA staff in Washington, with staff from EPA's regional offices partly responsible for and supporting the ideas and compromises involved.

Another difference between the 3M project's stakeholder approach and the approaches taken by the other three projects was the lack in 3M's case of a unified stakeholder group that worked together as a whole and in subcommittees. Local residents and city officials of the town of Hutchinson were represented on a stakeholder group formed by 3M. Yet the 3M group was separate from PPC, which was MPCA's stakeholder; neither had as frequent or as intense meetings as the Intel, Merck, and Weyerhaeuser stakeholder groups, and they did not get nearly as involved in the details of drafting the FPA. To have done so would have entailed a commitment of time that few would have been willing or able to make (see, for example, a discussion of the time spent by Chandler residents on the Intel project in NAPA 1997). Instead, 3M's local stakeholders acted more as a consultative body to assure 3M that its XL project would have community support. The group wanted reassurance that the project would not bring increased health risks to residents but raised few objections. Because the company had announced to the community that it would be discontinuing its magnetic tape products but had pledged to maintain most of the jobs involved in their production, the local community was concerned about protecting jobs as well as public health.

As has been noted, MPCA had its own advisory stakeholder group, PPC, which was representative of statewide business, government, academic, and environmental advocacy organizations (but not of the Hutchinson community). MPCA's XL team members met often with PPC, 3M consulted periodically with its Hutchinson group, and 3M and MPCA staff worked extensively with one another. However, there were few meetings of MPCA, 3M, and PPC together, and no meetings that included all four groups. There were only two or three meetings between MPCA and PPC to discuss proposed drafts of the FPA that included a staff member from EPA Region 5 on the telephone. The collaborative process was a combination of parallel and linear interactions and information flow among the stakeholders. Unfortunately, the information flow to EPA's Region 5 was quite poor. Also, few attempts were made to reach out to national environmental groups, although PPC had representatives from two environmental organizations.

In comparison with the collaborative processes engaged in by Intel, Merck, and Weyerhaeuser, the 3M process was less inclusive and less parallel. Among the stakeholders that were included in 3M's case, however, the less inclusive, semilinear approach made reaching a consensus agreement among local participants easier. Its fatal weakness, however, was that those excluded—EPA Region 5, to a lesser extent EPA headquarters, and to an even lesser extent nongovernmental organizations—built up distrust of and skepticism about the Minnesota group's draft FPA.[12] The Minnesota players had reached an agreement, but no one at EPA had been in on it or understood the bargaining process and the concessions, often tacit ones, that had been made.

Consequently, unlike the other three comparison cases, the project had no champion from the EPA regional office to explain and defend the draft FPA to others at EPA as well as to those outside the agency. Thus, these process issues along with the substantive issues mentioned earlier were the main causes of the failure to reach an agreement in the 3M case. If negotiating parties are able to make guarantees about results and to engage in a more inclusive decisionmaking process, there is a better chance for a cooperative solution. But making guarantees and working inclusively are not easy or possible in every instance.[13]

Notes

[1]The fact that they considered exiting and that EPA knew of this possibility actually may have contributed to their reaching agreement. When 3M threatened to walk away from the Hutchinson deal, EPA took up the negotiations with a new urgency that it had not demonstrated previously (Chapter 7).

[2]The criteria pollutant caps for the four major sources, which already were regulated to best available control technology standards, were fixed 43% higher than 1995 levels to allow for production at full capacity. However, these caps provided the facility no operating flexibility and simply quantified what current regulations allowed. The VOC cap for emissions from the plant excluding the small amount coming from the four major sources was set at the 1995 actual emissions level so that production increases could only be accommodated by proportionate decreases in VOC emissions per unit of production.

[3]Merck's XL permit did allow VOC emissions to increase substantially, although the agreement provided incentives for Merck to minimize these increases. Objections to these potential increases in VOC emissions were raised by several stakeholders and had the area not been designated nitrogen oxide–limited, the agreement might not have been approved in its present form. In the end, the plant's conversion from coal-fired to natural gas–fired boilers reduced criteria pollutants to such a low level that Merck had considerable bargaining power to negotiate the best deal possible. Had Merck already converted to natural gas (as had 3M), there would not have been an XL project at its Stonewall Plant.

[4]One might say that Intel, Merck, and Weyerhaeuser followed a route that involved "influentials first" (the EPA regions) and that tried to include "nearly everyone" (a broad coalition), whereas 3M and MPCA followed a route that involved "friends first" (MPCA) and did not attempt to include everyone at the first stage in the negotiations. By combining large numbers (a broad stakeholder coalition) and an "influentials first" approach (including the EPA region from the beginning), Intel, Merck, and Weyerhaeuser maneuvered better around a dilemma. A small-numbers

approach results in lower levels of cooperation, less equal outcomes, and a decrease in integrative trade-offs (Bazerman et al. 2000, 2001; Palmer and Thompson 1995). Including everyone first, conversely, increases complexity (Kramer 1991), which in turn leads to simplifying strategies and cognitive mistakes (Bazerman et. al. 2000). There also is the role that professional facilitators played. National environmental groups were not happy with any of the proposed agreements, including those with Intel, Merck, and Weyerhaeuser. However, professionally trained negotiators were at meetings EPA organized in which there was an airing of the issues in a relatively stress-free environment (a process that did not occur in the 3M case), thus suggesting that facilitators, schooled in the negotiations literature, can play a useful role.

[5]Productive technology in the semiconductor industry changed rapidly with each product cycle. Consequently, Intel was constructing new facilities that, in addition to having no operating history, would not be entirely comparable to microprocessor fabricating plants elsewhere.

[6]Water discharge permits typically included requirements per unit of production, quantities that were well established in the wood pulp and fluff-producing industry. The permit, though, also has absolute discharge limits.

[7]Several factors greatly complicated the judgement of whether 3M's proposed FPA would result in SEP. The manufacture of magnetic tape would be phased out and replaced by the production of various adhesive tape products, some of which were new. Adhesive tape and magnetic tape production had quite different pollution prevention and pollution control characteristics. Sticky tape production had more technological options for pollution prevention solutions than did magnetic tape production, and the solvents used, where necessary, contained fewer hazardous air pollutants (HAPs). Although magnetic tape production was being phased out, it would soon come under new maximum achievable control technology (MACT) standards that restricted HAP emissions. Several emissions units were not controlled and would require expensive controls under MACT. In addition to the several unregulated lines in the audiovisual tape operations, the adhesive tape producing facility also had many emissions units that were exempt (grandfathered) from Clean Air Act requirements as long as no major changes were made. Most of these units also had voluntary controls, but not all were necessarily up to the standards imposed by regulations. Eventually, EPA would also introduce MACT standards for adhesive tape production units, but it was uncertain when this would occur. Therefore, it was impossible to predict what the "otherwise applicable regulations" would require for the grandfathered units over the life of the XL permit.

[8]The permit and FPA had to accommodate 3M's plans for increasing output at the facility by up to 70% over 10 years. Unless emissions per unit of production were reduced proportionately, the production increase would result in increased emissions of VOCs and HAPs, increases that would be permitted under conventional regulations as long as all regulated units were satisfying Clean Air Act requirements. This likelihood that air emissions would increase over 10 years necessitated facilitywide emissions limits for VOCs and HAPs well above current actual emissions, unlike most of the caps in the Merck pilot and even the caps in the Weyerhaeuser pilot, which were at worst only 40–43% greater than actuals.

[9]Setting the 3M caps in units of emissions per unit of production was very difficult to do for the Hutchinson facility because it manufactured such a wide variety of tape

products. For instance, the length of audiovisual tape produced daily was orders of magnitude longer than that of Magic Tape, which in turn had characteristics quite different from Post-It Notes, and so on.

[10]Region 3 participants drove from Philadelphia to meetings in Virginia; Region 9 representatives flew from San Francisco to attend meetings near Phoenix. Ironically, the regional office closest to the site of its XL project, Region 4 in Atlanta, citing budgetary constraints, held meetings in Atlanta, roughly a two-hour drive for all the other stakeholders. Differences could be bridged face-to-face that often could not be bridged by teleconference. Nonetheless, no matter how inclusive these projects were, they could not include everyone. The Natural Resources Defense Council, for instance, was not a direct party to the negotiations in the Intel, Weyerhaeuser, and Merck cases; having been left out, it later came to oppose the proposed agreements.

[11]An exception was that a representative from EPA's Office of Air Quality Planning and Standards in North Carolina did participate in the Merck and Weyerhaeuser stakeholder group deliberations and contributed to their success.

[12]In addition, the Minnesota stakeholders remained unaware that their vision of an XL project was too radical for EPA staff. Would a more modest proposal—assuming that it was worthwhile to do and was acceptable to EPA—have been forthcoming, assuming that there was this awareness that EPA would not approve a more ambitious project?

[13]They also might not be desirable because they make the resulting agreement less experimental than it otherwise might be.

10

Roadblocks to Cooperative Solutions

In the four cases examined in this book, the companies wanted to deal in a creative way with significant environmental issues. They wanted to avoid what they considered complex, cumbersome, and bureaucratic pollution control rules. They participated in Project XL because they hoped to get their products to the market faster, considered the rules a straitjacket, and were looking for greater flexibility. The U.S. Environmental Protection Agency (EPA), for its part, sought better environmental performance than it considered possible under current and anticipated rules. To achieve this type of superior environmental performance (SEP), EPA offered regulatory flexibility. Many factors determined the outcomes of the cases in this book. The major roadblocks to achieving cooperative solutions upon which we have focused were (1) difficulties in establishing criteria for evaluating environmental benefits and balancing them against regulatory flexibility and (2) difficulties in establishing decisionmaking processes for making this type of determination (Stewart 2001). At its heart, the basic bargain, the quid pro quo, was problematic.

In this chapter, we examine real-life obstacles to using negotiated environmental agreements such as Project XL to achieve cooperative solutions to complex and contentious environmental issues. We provide conclusions from our analysis of the Intel Corporation, Merck & Company, 3M Company, and Weyerhaeuser Company cases, focusing on some of the more important intrinsic barriers. In the forefront of these obstacles, we argue, are issues of substance and process. But perceptions, influenced by the institutional roles of the participants, also play a significant part.

We start with a discussion of these perceptions. Why do negotiations seeking solutions break down? A prominent approach in the literature emphasizes perceptual biases that lead negotiators to want to "beat" the other side and "win" at any cost (Hoffman et al. 2002). These biases can prevent them from achieving win–win outcomes. Hoffman and others write:

> Negotiators representing environmental and economic interests often reach solutions that are not on the efficient frontier because of the assumption that they have opposing interests.... The mythical fixed pie prevents disputants from cooperating to integrate their interests.... This is a false assumption in virtually all complex negotiations.... The mythical fixed pie results from the tendency of people to overgeneralize purely competitive situations instead of seeing them as mixed motive situations.... Creating the trades to overcome the mythical fixed pie is exacerbated by an (often false) perception that issues in environmental negotiations are sacred. (Hoffman et al. 2002, 830–31)

This approach makes an important point. We, however, hold that it is not just biases in the way negotiators think that prevents agreement but the proposed terms of an agreement (the substance) and the process of how decisions are made. To this list of the barriers, we add the institutional roles that the participants play. These roadblocks—perceptual biases, substance and process, and institutions—are discussed in this chapter.

Perceptual Biases

Neale and Bazerman maintain that "situations are most useful if viewed from an interpretive perspective" (1991, 7), and that it is not the "objective external aspects of the situation" that are important, but how "the negotiator perceives the situation and uses perceptions to interpret and screen information." According to these analysts, the judgements of negotiators are systematically distorted by their biases, including: incorrect framing of risks; anchoring points of view in irrelevant information; and overreliance on readily available options. Neale and Bazerman believe that negotiators use stereotypes, schemas, scripts, and other mental shortcuts to cognitively frame disputes (Weick 1979). In failing to consider their opponents' perspectives, they often devalue the real concessions their opponents are willing to make.

Neale and Bazerman (1991) argue that structural factors are less relevant than perceptions because these factors are beyond the participants' control and the participants cannot do much about them. The challenge is to change negotiators' perceptions by making them more "rational," and the way to do so is by giving them enhanced cognitive skills.[1] Given that the core problem is a psychological one and that the emotions of the parties tend to overwhelm their logic, the solution is to change the parties' perceptions (Susskind and Cruikshank 1987). If at the start of a negotiation, they do not trust each other, then trust can be built when each side—often with the help of a trained facilitator—learns to understand the other side's perspective. The onus is on the parties to the negotiation to redefine, recast, and reframe the problems they confront so that what seem like win–lose outcomes are transformed into win–win results (Lewicki and Litterer 1985).

Inspired by the popular book *Getting to Yes* (Fisher et al. 1991), analysts have focused on how to create a positive environment, where negotiators trust one another and are willing to collaborate. This point of view is optimistic (Fisher 1994). The way to improve the process, get it unstuck, and handle conflict in a more constructive way is for each side "to speak in another's voice," "step into their shoes," "feel empathetically" how they feel, and develop comprehension for their positions without necessarily agreeing with them. Freshly liberated from past constraints, the parties should engage in a vigorous search for their shared interests. They should brainstorm, be open to wild ideas, and break free from the factors that prevent them from forging agreements (Fisher 1994). The advice that emanates from the perceptual school is that people should identify biases in how they make decisions, develop a rational outlook, break the fixed-pie mentality, search for trades that improve everyone's lot, and consider the effects of their decisions on future generations (Bazerman et al. 2001).

Game theory's approach is different, but some of the implications are similar (Raiffa 1982; Axelrod 1984). Repeated interactions are needed to forge cooperative solutions. Thus, the parties have to observe the other sides' behavior and work with them. Through trial and error, the parties can discover if mutual rewards are possible.[2] For repeated interactions to take place, there has to be a positive atmosphere for bargaining. If the conditions for mutual gain are not ripe, because "inadvertent careless friction" festers and spoils "the atmosphere for bargaining" (Raiffa 1982, 13), then there is no alternative but to abandon the negotiations. If one side or the other feels it is dealing with "strident antagonists, malevolent, untrustworthy characters whose promises are suspect" (18), then it is unlikely that an agreement can be reached.

Substance and Process

The mutual-gains school of negotiation holds that people can be led to revise their perceptions. This insight is valuable and well intentioned, but in the cases we examined the participants did not experience difficulties just because of an inability to overcome perceptual biases. Though these biases may have played a role, other factors were involved. Substance and process

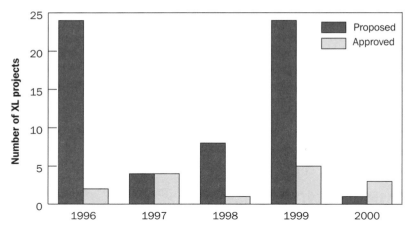

Figure 10-1. Project XL Proposals and Approvals, 1996–2000

Note: See Hoffman et al. 2002, 827. After August 1999, EPA did pick up the pace of approvals, but of the 36 XL agreements reached after that date, 23 were with government organizations or nonprofits and only 13 were with businesses (also see Stewart 2001). By the end of the Clinton administration, EPA had approved 50 XL pilots.

and not just cognition prevented the parties from achieving cooperative solutions. In this book, we have emphasized the importance of substance and process as barriers to cooperative solutions.[3] The interplay of these factors does not necessarily mean that agreement is not possible; rather, as our cases show, the process of reaching agreement often is painful, frustrating, and time consuming, which puts a limit on how useful cooperative approaches can be. The purpose of Project XL—50 negotiated environmental agreements within a year—was to test ways of producing SEP with improved economic efficiencies while increasing public participation (Clinton and Gore 1995; U.S. EPA 1995b). However, far fewer agreements were negotiated in this time frame than anticipated (Figure 10-1).

Despite laudable goals, lofty ambitions, and good intentions, efforts to find common ground among businesses, governments, environmental groups, and communities have been beset by difficulties. Uncertainties and inconsistencies have resulted in high transaction costs, which have deterred applicants from the business community from coming forward with interesting ideas (Blackman and Mazurek 1999; Stewart 2001; Delmas and Mazurek 2001). Programs like Project XL have been plagued by low industry participation. Although the promise of cooperative agreements is high, the practical impediments to reaching them have been great.

The substantive and process barriers to cooperative solutions we have highlighted in this book include the trade-off between SEP and regulatory flexibility, the different goals the parties have, and technical, economic, and legal uncertainties. Given these uncertainties, the parties had difficulties in deciding on an appropriate negotiating group and negotiating process. They had trouble drafting an initial accord and securing support for it.[4] Starting with the quid pro quo we briefly discuss these barriers.

The Quid Pro Quo

The main substantive barrier discussed in this book is the quid pro quo—in exchange for SEP, a facility seeking a permit is to be granted regulatory flexibility. The problem with this framework is how to define SEP and how to work out trades between different levels of SEP and different levels of flexibility (see Chapter 2). The quid pro quo was a highly imperfect means for reaching cooperative agreements. Granting regulatory relief and thereby operating flexibility provides a company with tangible economic benefits. These benefits must be balanced by environmental gains that would not be otherwise achieved. Because the flexibility has to be "paid for" with superior performance, a company must show that the environmental benefits it offers otherwise would not have been gained. This criterion can be difficult to satisfy. Because definitions of SEP and flexibility are vague, companies considering participation in XL are presented with a level of risk that is hard for them to accept.[5]

The quid pro quo, by focusing attention on guaranteed benefits, limits learning. Because process and substance are interconnected, it significantly slows down the negotiations, creating a bargaining atmosphere where each side strives for the best deal. By making it easier for dissenters to raise objections that the SEP offered is not good enough, it stands in the way of deals being consummated. Therefore, it leads to obstruction and high transaction costs that prevent an approach like Project XL from being used on a widespread basis. The framework for reaching an agreement under XL had serious limitations that plagued efforts to reach cooperative solutions under this program.

Different Goals

That the parties trying to reach an agreement have different goals is another serious obstacle to cooperative solutions (Chapter 3). In the negotiations literature, a distinction is made between positions and interests (Susskind et al. 2000). The literature holds that if the parties to a negotiation can advance beyond the respective positions they profess and address their deeply held fundamental interests, they are more likely to craft a mutually beneficial agreement. However, before the parties try to reach an agreement, they are unlikely to have either well-articulated positions or a good understanding of their interests. They are more likely to have vague goals that reflect their interests (see Chapter 3). These goals are a result of their history and a complex set of expectations that reflect the relationships the parties previously have had. These goals are not necessarily found in explicit form in policy documents and statements, but rather can be inferred from the parties' orientations, actions, and behavior. Complex combinations of worldviews, imperatives, incentives, the practical results of these goals can be seen in what the parties do—both in how they make decisions and how they aim to legitimize their decisions in the eyes of others.

The negotiators in Minnesota were able to work out a common front only when the participants agreed they could achieve environmental and economic benefits by doing more than regulations required. Their goal was an innovative pilot that would change the system, but EPA saw such change as risky. Along with national environmental organizations, it was reluctant to conduct experiments that might jeopardize the status quo. That business and government have different goals is well recognized. EPA wants to protect and improve public health and the environment, and businesses want to enhance their competitiveness and lower their costs. However, the environmental managers participating in XL from Intel, Merck, 3M, and Weyerhaeuser also wanted to improve environmental performance within the limits imposed by economic considerations. Though government regulators argued that economic considerations were of no concern to them (recall the comments of David Ullrich, deputy administrator of EPA Region 5, at the May 1996 meeting in Chicago), they had to recognize that these considerations were important as some of the regulations they administered called for technically feasible outcomes at a reasonable cost. Notwithstanding the fact that business managers' and regulators' goals overlap to some degree, they still found it hard to agree.

It is not unusual for parties to a negotiation to have goals that are not perfectly aligned. Typically, though, the parties are not monolithic. Some of EPA's staff—for instance, those in enforcement—felt more threatened by the government's XL initiatives than others. Indeed, sources of joint gain may arise from these differences in goals. The different valuations allow for intricate trade-offs; stable agreements emerge as each party feels that some of its goals are satisfied (Lax and Sebinius 1986). The Weyerhaeuser case shows the benefits of multifaceted trade-offs. Water, air, and solid waste issues were pertinent. The promise to try to make a fundamental technological breakthrough that EPA could use as a standard at other facilities helped to clinch the deal. In contrast, the 3M case mostly boiled down to a discussion about VOC emissions; as a consequence, there was less room for these trade-offs.

Nonetheless, one of the most important reasons Project XL ran into trouble was because EPA staff did not have clear goals. Without such goals, it was difficult for them to manage the relationships they had to develop with businesses and environmental groups. The different goals of the EPA offices constrained staff who wished to pursue XL agreements more vigorously. As pointed out by Wood and Colosi (1996), in a negotiation, there can be as many as four or more subnegotiations taking place at the same time. EPA staff working on Project XL had to deal with their own negotiating teams; negotiating teams in other parts of the agency (such as the regions, the air and water programs, the enforcement office, the general counsel); outside negotiators (for example, 3M, the Minnesota Pollution Control Agency, and the Pilot Project Committee); and stakeholders (such as the Natural Resources Defense Council). Under such circumstances, it was difficult to make trade-offs. It was also difficult to exchange less important objectives for more important ones and come to agreement.

Technical, Economic, and Legal Issues

Complicating factors—technical, economic, and legal uncertainties—also were obstacles to agreement (see Chapters 4 and 8). The complex technical questions at 3M's Hutchinson, Minnesota, plant were exacerbated by the uncertainty generated by a corporate restructuring that came about because of changes in 3M's economic circumstances. EPA, the Minnesota Pollution Control Agency (MPCA), and 3M proceeded without knowing if an agreement could be carried out without violating environmental laws and requirements. A related matter was MPCA's relationship with EPA Region 5: Did MPCA have the authority to issue an XL permit on its own? Technical, economic, and legal uncertainties such as these complicate matters and raise questions—for instance: When and under what circumstances will a new, more environmentally advanced technology be ready for commercial application? What if a company suffers a business reversal in the middle of a negotiation and has to sell off assets or shut down production? If the legal foundation for an agreement is weak, what legal principles can be used to fashion one?

All of these factors play an important role. Environmental management often requires sophisticated technical answers. Because pollution prevention approaches often involve new, unproven techniques, the more an experimental pilot seeks to explore innovative solutions, the greater is the risk of failure. In the 3M case, economic circumstances imposed barriers that did not arise in the Intel, Merck, or Weyerhaeuser pilots; nonetheless, the high transaction costs of negotiating an agreement were an important obstacle in all four cases. Uncertainty about the legal flexibility in environmental statutes also discourages experimentation. Legal issues were significant contributors to the high transaction costs. Neither business nor government wants to be bogged down in expensive, time-consuming litigation.

The Process of Getting Agreement

The process for reaching agreement proved to be a long one in each instance. The parties had to create a provisional accord (discussed in Chapters 5 and

8) and obtain support for it from other parties (discussed in Chapters 6 and 8). In the 3M case, they reworked the initial accord and started again with new actors playing significant roles (discussed in Chapter 7). Handing off a proposed agreement from group to group slows the process and creates misunderstandings. From the start, the Intel, Merck, and Weyerhaeuser cases involved a more parallel, inclusive process than the 3M case (discussed in Chapter 9)—an important reason the participants in those cases were able to reach agreement, whereas the 3M participants were not. Nevertheless, even in cases where agreement was reached, once the approval process went beyond a preliminary accord, it became bogged down in new controversies. Additional compromises had to be worked out and the rationale for earlier decisions revisited.

The preliminary accords depended on an intimate understanding of site-specific issues. Yet, the right to final approval was in hands of EPA headquarters. National environmental groups such as the Natural Resources Defense Council were in contact with EPA headquarters about XL pilots but not with the EPA regional offices that were involved in individual project negotiations. Consequently, neither EPA headquarters staff nor the environmental leaders were privy to early discussions. Some of the participants, who had lost arguments at an earlier point, tried to reopen issues at a later one. They appealed to EPA headquarters staff to give their views a second hearing.

Thus, parties trying to create an agreement not only have to create agreement among themselves. They have to obtain additional support from other parties, who are often more influential and perhaps were not involved in the original negotiations but who nonetheless must give their backing if the project is to go forward. Obtaining support from parties beyond those who formulate an initial agreement is an obstacle to reaching cooperative solutions. Negotiating an agreement is not a one-time event. The negotiations are not conducted and concluded once but must be repeated again and again as more groups are brought into the process and their support must be won.

To summarize, negotiations for cooperative environmental solutions break down not just because of biases in the way negotiators think, but more importantly because of the actual terms of the agreement (the substance) and the process of how decisions are made. In the cases we examined (in Chapters 7 and 8) with respect to substance, there was a better chance for agreement if the parties were able to guarantee the results. With respect to process, there was better chance for agreement if the parties engaged in a more inclusive decisionmaking process. Two factors differentiate the approved Intel, Merck, and Weyerhaeuser projects from 3M's: first, the near-certain SEP offered in the three approved projects, in contrast to the uncertainties in the 3M case and, second, the negotiating process itself, most notably the broader coalition in the three approved projects as opposed to the narrower one in the 3M case.

Institutions

While issues of substance and process are in the forefront of the obstacles to cooperative solutions, institutions and the way they influence perceptions play a significant role. When people act as representatives of organizations and other institutions, they behave differently than when they act on their own. In addition to the substantive and process issues we have discussed, institutional barriers exist. They interact with substance and process, making agreement difficult to reach.

Psychologists, social psychologists, game theorists, and decisionmaking theorists tend to view the barriers to reaching agreement in psychological and perceptual terms. They maintain that to succeed in a negotiation the parties should separate people from the problem, focus on interests and not positions, generate a wide range of options for reconciling opposing points of view, make sure that these options have something in them for both sides, and insist that the solutions should be based on objective criteria (see for example Susskind and Cruikshank 1987; Neale and Bazerman 1991; Fisher 1994). Other analysts—sociologists, political scientists, organization theorists, and system theorists—tend to see the difficulties in how institutions customarily relate (Dougherty and Pfaltzgraff 1970).

The nineteenth-century French sociologist Emile Durkheim wrote that "every time that a social phenomenon is directly explained as a psychic phenomenon, one may be sure that the explanation is false" (as quoted by Dougherty and Pfaltzgraff 1970, 140–41). Durkheim overstated this argument but we agree with him when he wrote:

> Social facts do not differ from psychological facts in quality only: They have a different substratum; they evolve in a different milieu; they depend on different conditions.... The mentality of groups is not the same as that of individuals. (as quoted in Dougherty and Pfaltzgraff 1970, 140–41)

The problem with the psychological approach, though it has obvious attractive features, is that it presupposes that institutional issues can be overcome by changing cognition. However, the reality is that not every attempt to bring about a dialogue among opponents results in a magical transformation in which contending sides suspend belief in their preconceptions and achieve new respect for the value of the other side's positions.

It is good to keep in mind that excessive idealism, founded on the assumption that people are naturally inclined to cooperate, can be naive, and that a belief in inherent goodwill that shifts the focus from what real people actually do to what reasonable people should do misses the point that some environmental disputes are nearly intractable. The rhetoric is so heated, the emotions so high, and the characterization of the parties so stereotyped that a negotiated agreement is not likely (Gray 2000).[6] Recent research (Conlon and Sullivan 1999; Brodt and Dietz 1999), therefore, has started to move from a heavy emphasis on cognitive barriers to an emphasis on institutional factors. Institutions play an important role; outside the minds of the negotiators, a social reality exists, which has a life of its own and which can present additional barriers to reaching an agreement.

According to Hoffman and others, the literature on institutions has three pillars, regulative, normative, and cognitive:

> Regulative aspects of institutions are based upon legal sanction to which organizations accede for reasons of expedience. Normative aspects of institutions are morally grounded; organizations comply based on social obligation. Cognitive aspects of institutions reference the collective constructions of social reality via values, language, meaning systems, and other rules of classification embodied in public activity.... These three aspects are operationally intertwined ... and are present in all forms of institutional control. (Hoffman et al. 2002, 833)

Institutions, these analysts maintain, affect what people do through "laws, rules, protocols, standard operating procedures, and accepted norms" (Hoffman et al. 2002, 833).

Some of the institutional barriers we observed in the cases discussed in this book were based on laws, rules, protocols, standard operating procedures, and accepted norms. Our list of barriers does not include every issue, nor is it our intent to review all the literature that might have bearing. Some of our ideas about the institutional barriers come from Wilson, who celebrates the U.S. system but also criticizes its tendencies, writing that government agencies are part of a fragmented and open political system, which "is not designed to be efficient ... but tolerable and malleable" (Wilson 1989, 376). We discuss each of these institutional barriers in turn: a reluctance to take on controversial precedents; a tendency to treat similar cases alike; the exercise of discretion creating uncertainty; an ingrained culture of skepticism and mistrust; a perception of insufficient legal authority; and gains made from inefficient policies. Issues about how institutions such as regulatory agencies customarily relate to those they regulate emerged from our analysis, but this fact does not negate the importance of institutional barriers on the business side, such as the paramount significance of economic concerns, an inability to tolerate government inefficiencies, and a failure to completely recognize the importance of competing but equally compelling goals in government.

A Reluctance To Take On Controversial Precedents

Officials of regulatory agencies such as EPA often are loath to take on controversial precedents that might arise in potentially ambitious programs such as Project XL (Wilson 1989). The most common changes that they are willing to pursue are "add-ons" (Wilson 1989), similar in nature to the series of reforms and experiments to which EPA has adapted in the past. Add-ons do not challenge an agency's basic operating philosophy or principles. Innovations in government agencies are rarely about their core tasks. Typically, they are on the periphery. The tendency to treat innovations as peripheral is especially true when the innovations are externally imposed. Indeed, although EPA participated in formulating the White House's regulatory reinvention initiatives, the White House, reacting to a Congress suddenly controlled by a Republican majority, basically imposed XL on EPA. It was not surprising, therefore, that regulatory reinvention was not central to EPA's priorities.

Like other reforms in EPA's history, Project XL never overcame its marginal status. In comparison with the old-style command-and-control activity in which EPA was engaged, negotiating agreements with business was not an important activity. For example, some EPA staff members dismissed XL as not much more than "election year window dressing" (Susskind and Secunda, 1999, 100). Susskind and Secunda quoted one "high ranking EPA director" as saying that Carol Browner's leadership style was "to survive short-term political drills, not to advance policy" (1999, 103). The EPA culture was one in which failure was not tolerated, and thus experimentation was hard to carry out (Susskind and Secunda 1999, 98–99).

There were additional reasons why many EPA staff members may have been reluctant to take on the challenge of a project like XL. They may not have been fully supportive of efforts to reach cooperative solutions (O'Leary 1995) because they believed they would have less control under cooperative agreements than they did under conventional command-and-control regulation. They may have been concerned about powerful business interests compromising aggressive antipollution efforts (Freeman 1997). And they may have been worried about the costs to their programs as cooperative solutions stretched their resources.

In addition, the way EPA delegates authority to the states and regions tended to reinforce XL's marginalization. EPA delegates to states the authority to issue and enforce permits. In turn, the agency's regional offices review state-issued permits and enforcement actions and oversee state actions. The regional offices focus on existing rules and regulations while headquarters is responsible for new policy and developing new rules and regulations in response to legislation passed by Congress. Given this division of labor within the agency, the regional offices are not encouraged to pursue innovative changes, even though their staff is closest to the facilities that EPA regulates.

A Tendency To Treat Similar Cases Alike

In government agencies such as EPA, there also is a conflict between equity (treating everyone the same) and responsiveness (tailoring responses to specific settings and circumstances). This conflict is between treating situations based on clear rules known in advance and taking into account special needs and conditions (Wilson 1989). Under Project XL, EPA was trying to move from treating cases alike toward greater responsiveness. The Aspen Institute proposals, 3M's Beyond Compliance Bill, and the Bill Clinton–Al Gore initiative aimed to create two systems: For most companies, the system would be based on a uniform set of rules. But for companies that had gone beyond compliance, there would be more flexibility. The purpose of treating some companies differently was to achieve greater environmental and economic benefits.[7]

Some government agencies can follow the model of deciding each case on its merits. However, agencies in the health and safety area, including EPA, the Occupational Safety and Health Administration, and the Food and Drug Administration, tend to adhere to rules absolutely (Wilson 1989). Conscientious officials consider it baleful to break the rules when their purpose is to protect human life and ecosystem integrity. Typically, staff members of government agencies seek the comfort and stability of operating according to rules (Wilson 1989; also see Orts 1995a, 1995b). Their bias is to solve problems by adopting rules, which leads them to treat all cases alike even when there are good reasons not to do so.

The Exercise of Discretion Creating Uncertainty

Rules—whether to guarantee equity or to assure responsiveness—are generally unwieldy, partial, and conflicting (Wilson 1989). Therefore, discretion is inevitable. And when there is discretion, organized interests—such as businesses and environmental groups—seize it and try to use it for their purposes. The rules are meant to insulate government officials and keep them from exercising discretion, but there is a limit to how far the rules go. They are never complete and comprehensive enough to cover all situations. Although a relationship exists among statute, regulation, and internal procedures, this relationship is by no means deterministic. An agency's staff interprets laws and establishes regulations and operating procedures that govern behavior. It can administer the laws Congress has passed in a number of ways. If the agency creates additional uncertainty in administering these laws, it leads to higher transaction costs, a problem that plagued EPA in carrying out XL (Blackman and Mazurek 1999).

Rules alter human relations in ways that lessen or increase the motivation to negotiate. If the rules are ambiguous, they can increase tension among the parties trying to negotiate. If the rules fail to take into account cost and feasibility, businesses will not take them seriously. But if the rules do not maintain high standards of environmental performance, public interest groups are likely to be mistrustful, lest businesses be relieved of the burden of what the law intends: protecting the public from harm. Though talented, motivated people can get around the rules when they do not make sense, typically such people are in short supply (Wilson 1989). In Project XL, whether they were in short supply was not as important as the incentive and reward structure of EPA's consensus style of decisionmaking, which made it difficult for such people to buck the system. It was difficult for them to create workable definitions of concepts like SEP and institute straightforward processes for resolving disputes about SEP, flexibility, and other issues (see Chapter 11).

For the parties that tried to reach cooperative solutions under XL, it was hard to deal with EPA because it was not clear how the agency would define SEP and how it would resolve the trade-off between SEP and flexibility. Because key terms like SEP and flexibility were not well defined, it was difficult to reach negotiated agreements under this framework. It was inevitable that intense bargaining and resulting conflict would arise over just how much SEP was needed to justify the economic benefits derived from a given amount of operating flexibility. In the end, people in the business community were unwilling to engage in this type of protracted bargaining. Dave Sonstegard, head of the Environmental Technology and Safety Services Division at 3M, believed that a more streamlined system was needed to revitalize XL. He argued for:

- a clearer definition of the role of superior environmental performance so that companies would understand the ground rules better;
- consistent expectations from players in the EPA and outside the agency, including state agencies, national environmental groups, local stakeholders, and other interests;
- a distinct definition of the rules, including who was to be involved and what decisionmaking authority they would have;
- high-level policy guidance and decisionmaking from a team of top EPA officials who would work closely with companies, state governments, and stakeholders; and
- a crisp management structure among staff reporting to the top officials so that delay would be avoided.[8]

To stretch and probe for reform, companies like 3M were seeking less uncertainty than EPA was able or willing to offer.

An Ingrained Culture of Skepticism and Mistrust

Our interviews with EPA staff participating in the 3M and other XL projects exposed an ingrained culture of skepticism and mistrust toward industry motives. Several were quite explicit about it. They maintained, for instance, that 3M was trying to take advantage of its previous voluntary controls to be able to increase at will its emissions in the future. This attitude led to strong concerns at EPA over the enforceability of XL agreements, concerns that greatly complicated the negotiations with 3M, as well as those with Merck and Weyerhaeuser. Indeed, negotiators from EPA and Merck spent months getting the permit right after a basic agreement had been reached.

Susskind and Secunda have singled out EPA's culture of confrontation as a barrier to the agency moving to a more collaborative approach. According to Susskind and Secunda, "EPA's institutionalized 'enforcement culture' is a critical barrier to operating in the new ways called for by Project XL" (1999, 95). They quote one EPA administrator, "Philosophically, the greatest resistance to flexibility is within the agency itself" (1999, 96). EPA's concerns arose out of an inherent suspicion that business would seek to manipulate the system without really benefiting the environment. EPA staff members who participated in XL were considered "turncoats" by some managers (Susskind and Secunda 1999, 97), and sufficient resources were not awarded to carry out XL projects. The traditional hostility between EPA and the regulated business community made it difficult to communicate and form teams with business.

A Perception of Insufficient Legal Authority

EPA staff, perhaps because of the culture of confrontation, operated under the assumption that they had insufficient legal authority to carry out Project XL. According to a 3M team leader, EPA's Office of Enforcement and Compliance Assurance made it clear in an early memo that XL projects could result in the violation of legal and regulatory requirements.[9] The initial answer EPA gave as to how it would proceed under XL was enforcement discretion, a route typically taken with violators when they are under an enforcement agreement to move toward compliance. Under enforcement discretion, an XL participant was particularly vulnerable to citizen lawsuits.

This legal strategy was not acceptable to business. EPA therefore decided to switch to site-specific rulemaking as its preferred method of doing XL projects (U.S. EPA 1997a, 1997f). However, the transaction costs of site-specific rulemaking typically were high. In this instance, EPA had to show how the terms of a project were an alternative means of complying with existing law. It had to issue a formal rule, publish it in the *Code of Federal Regulations*, and receive formal comments. The notice and comment method could take as long as six months to a year. As a result, the total costs of a project might be greater than the benefits to the business participants. Though EPA had other legal routes available, because of its culture and standard operating procedures it chose not to take them.

Hirsch (1998) has explored a number of legal routes EPA could have taken under existing legislation, including one called "implied waiver authority," which would have given EPA more discretion to conduct Project XL-like experiments. To begin with, EPA simply could *interpret its rules in a flexible* way. All it had to provide was a written explanation, which it would make publicly available. However, flexible interpretation historically had been limited to statutory or regulatory language, which was unclear or ambiguous. Another possibility was that under some laws that EPA administers (for instance, Section 111 of the Clean Air Act), a company could *petition for a waiver* if it was using innovative technologies and needed flexibility. However, even under these laws the grant of a waiver would be subject to court review of whether EPA had acted in an arbitrary or capricious manner.

The limitations of these methods led Hirsch to conclude that there was a better solution, which was to rely on *implied waiver authority*. The theory of an implied waiver was that in any statute or regulation the drafters did not intend those who implement the law to go against the law's very purposes or cause undue or unintended hardship. Thus, the implementers had leeway to create flexible alternatives in special circumstances. These special circumstances were unintended and undue hardship on a particular party, disproportionate and unintended share of the burden, gross inefficiencies, manifest subversion of the law's aim, and piloting a new regulatory approach. Judicial review was not avoided, but costly notice and comment rulemaking was.

Hirsch admits that the best solution might have been for Congress to pass new legislation. He attributes the inability of Congress to pass new legislation to a combination of politics and the technical inadequacies of the legislation that was proposed. Senator Joseph Lieberman of Connecticut had introduced a bill (S. 2160, 104th Congress, 1996) that would have allowed a company to enter into an alternative compliance strategy provided that it could show "better overall environmental results than would be achieved under the current regulatory requirements." The Lieberman bill never made it out of the Senate Committee on the Environment and Public Works. There were technical problems with the bill, among them being that it put the burden of proof for showing SEP on EPA and required stakeholder participation but did not define the relevant stakeholders or provide a mechanism for their participation. Lieberman's approach continued to regard Project XL as a trade-off of SEP for flexibility, which limited the experiments that could be carried out.

Gains from Inefficient Policies

Environmental advocacy groups are by no means homogeneous in their opinions about and approaches to environmental protection. Nevertheless, taken as a whole, these groups constitute an institution with its own important perspective, which exerts considerable influence over environmental policies. Large environmental nongovernmental organizations did not have the resources to participate in individual XL deliberations. Environmental advocates who did participate usually represented local or regional organizations or were members of a local chapter of a large national group. Even so, the larger national organizations actively participated by commenting on pilot project proposals and agreements and by carrying their concerns to top EPA managers and other Washington leaders.

National environmental organizations regarded the 1994 election of a Republican-controlled Congress and the Contract with America as a summons to regulatory rollback and retrenchment (discussed in Chapter 2). Though some were willing to compromise with business to preserve environmental gains, others saw the Republican-controlled Congress as a threat, a sign that no compromises should be made and that increased cooperation with business should not be considered. The Clinton administration's response to Republican control of Congress was regulatory reinvention, of which Project XL was a part. The response of the environmental community to the administration's plan was not favorable (Skrzycki 1996).

Spence and Gopalakrishnan (1999) ascribe responsibility for Project XL's problems to the influence environmental groups had on EPA. It was not that national environmental groups did not value the environmental benefits that a program like Project XL could realize. Nor were they absolutely opposed on principle to win–win solutions. But they were playing a dynamic game, seeking broad improvements in environmental protection—not just now but for many years to come. Given that they were playing this long-term game, many national environmentalists believed their goals would be better achieved if status quo environmental policies were *more inefficient* rather than less. The more inefficient these policies were, the more industry, in the end, would be willing to make concessions and achieve ever higher levels of environmental protection to realize the efficiency gains it desperately needed. Spence and Gopalakrishnan use this reasoning to explain the tightening of formal Project XL criteria over time to place more emphasis on environmental benefits and less emphasis on efficiency. They write that "to the extent that individual firms or sets of firms can realize efficiency improvements through reform initiatives, national environmentalists lose … leverage over those firms" (Spence and Gopalakrishnan 1999, 28). They argue that the changes to the XL program discussed in the next chapter "strengthened the hand of national environmentalists in the XL bargaining process" (1999, 32).

Although Spence and Gopalakrishnan may overstate this argument, it does help explain why environmental nongovernmental organizations such as the Natural Resources Defense Council so persistently criticized XL agreements. Parts of the environmental community had not been happy with the widespread participation by environmentalists in the Aspen Dialogue and the Common Sense Initiative. They did not share the views of Environmental Defense, which, along with several other environmental organizations, had shown a greater receptivity to cooperation with business. Many in the environmental advocacy community saw Project XL only as a stopgap measure that the Clinton administration proposed to prevent a worse setback. They did not view it as the natural culmination of discussions earlier in the decade of how to work together with business to create a more sustainable future. Nor did they see it as an experiment in building new approaches to deal with the environmental issues of the future: global warming, biodiversity, overpopulation, and resource scarcity. They did not understand it as the next stage in the evolution of a new pragmatic approach to environmental management.

Instead, Project XL was seen by these environmentalists as an attempt to blunt Republican criticism of the environmental regulatory system by eliminating or reducing features the business community regarded as most onerous. For some, XL was tainted by its birth as a response to a political attack on the existing regulatory framework by a misguided Congress. For others, as long as the Republican threat seemed real, reform efforts like XL might be a reasonable gambit for the Clinton administration to pursue as long as EPA was able to extract from business a commitment to a higher level of environmental protection. With its ability to grant businesses and environmentalists some of what they wanted, XL was acceptable to them, but *only* if it was able to hold back the momentum for regulatory rollback. Still other environmentalists—those who were moving toward a more cooperative and less adversarial approach and had made considerable commitments of time and resources to the Common Sense Initiative (CSI)—were concerned that XL would provide a means for facilities to bypass agreements that might be reached through CSI. EPA Project XL leaders did not modify the project in response to the latter concerns. Early decisions by EPA about the stakeholder process in XL allowed for diverse, ad hoc arrangements for each pilot. Environmentalists appealed to EPA to make XL stakeholder participation more parallel to CSI. Three of the first eight XL participants were companies in the electronics industry, which was one of the six industries chosen for CSI. Environmentalists asked EPA to have CSI play an important role in all XL pilots involving facilities in the industries participating in CSI, a request that was denied (see Chapter 2).

During the 1996 presidential campaign, a key theme in Clinton's strategy for reelection became holding steadfast on issues of environmental protection against potential onslaughts by industry. From public opinion polls, politicians got the message that the American people would not tolerate the dismantling of 25 years of regulations designed to protect air, water, and other resources. Environmentalists and elements within the Democratic Party found Republican views about the environment in the Contract with America easy to attack. They saw less urgency for the kinds of compromises they believed they would have to make in XL and an opening to pursue an agenda of stricter regulatory enforcement and broader regulation. Rather than negotiating environmental agreements with business, advocacy groups believed that government should expand regulation to include health and ecological problems not adequately covered by current law. According to these groups, EPA should not emphasize flexibility to business but instead focus its efforts on new threats to the environment, such as nonpoint-source pollution, biopersistent compounds, and greenhouse gases. They saw an opening to pursue an agenda of stricter regulatory enforcement and broader regulation.

Intrinsic Problems

In sum, the cases in this book bring to light some of the real-life impediments to negotiating cooperative solutions. It is not just biases in the way the parties think that prevent them from agreeing, but more importantly the very substance of the accord and decisionmaking process for reaching agreement. Substance and process are serious challenges that can inhibit or facilitate agreement. Institutional issues also shape perceptions, affect substance and process, and remain obstacles to the diffusion of cooperative approaches. Because of factors such as these, it is probably clear to most people, including us, that negotiated agreements cannot be used to settle most environmental disputes. Conventional regulation, based on commanding and controlling businesses, has a place and no doubt will continue to be important in the near future. For now, we believe that the movement toward cooperative approaches should be in the form of further experiments, and in the next chapter we make some suggestions about how these experiments should be carried out.[10] Treating Project XL pilots as experiments to test new regulatory approaches is quite different from regarding them as precedent-setting alternatives to existing regulation. If individual pilots are treated as experiments, the stakes are lowered and the conditions for successful cooperative outcomes can be more easily established.

Notes

[1]Certainly, it makes sense for negotiators to be more "rational," but what does rationality actually mean? Are environmentalists irrational when they reject win–win solutions in individual cases in order to maintain their overall bargaining position in the long term (Spence and Gopalakrishnan 1999)?

[2]This approach is called "tit-for-tat."

[3]Prior research on alternative dispute resolution and negotiated rulemaking tends to support this conclusion. Bingham (1986), for instance, found that the "reason for the relatively high success rate" in alternative dispute resolution processes was that the mediators conducted "dispute assessments" about the *substance of the issues* (p. xxii) beforehand. They did not start the negotiation process if it was unlikely that a settlement could be reached. Bingham also considers process to be very important.

"Whether those with the authority to implement the decision participated directly in the process" (p. xxiv) helped to determine success. Because of substantive reasons, Stewart (2001) concludes that the use of negotiated rulemaking is limited. The rule being negotiated should not be too controversial. It must be easy to implement. Process also plays a part, as it must be easy to identify the stakeholders. The stakeholders have to agree negotiated rulemaking is their best option, feel they have something to gain, and be eager to see a resolution. Focusing on process, Coglianese (1997) shows how difficult reaching consensus is. Funk (1986) criticizes negotiated rulemaking because relevant interests are not represented. Rulemaking is therefore biased in favor of organized interests.

[4]When parties that historically have been at odds try to work together under these conditions, the negotiations can magnify their differences. The delays that occur can cause the process to unravel. The terms of the settlements may be so complicated that they are difficult to carry out. The solutions may be so diluted by compromise that they satisfy no one.

[5]Nor has EPA consistently applied this criterion for SEP in individual pilots. Of the four cases we discuss, only the Merck project could be said to provide environmental benefits that would not have been achieved with existing regulations. Stonewall's boiler fuel conversion would not have taken place without the regulatory benefits granted to the facility.

[6]There are three main schools in the negotiations literature (Arrow et al. 1995; Muthoo 1999). *Realists* would say that negotiations are about confrontation. Win or lose in nature, they involve zero-sum games and are distributionist in character. Their purpose is to distribute a fixed pie. *Idealists*, conversely, argue for cooperation, for breaking the mythical fixed-pie outcome and achieving win–win results. They advocate discovering mutually beneficial trade-offs that are integrationist in nature. *Pragmatists* are for expanding the pie as much as possible, but they recognize that there is an end as to how much the pie can be expanded and there will be conflict over what remains. It is worth developing this distinction a little further.

Realists see negotiations as a form of battle between groups with incompatible needs, opposing interests, and beliefs who have an emotional dislike of each other and distrust one another. The purpose of the parties is to maximize their power and dominance over the other sides. Their task is to know their own sources of power—resources, statutory, authority, expertise, and control of the agenda—and try to limit their opponents' sources of power. The advice that realists would give is *not to negotiate* if it is possible to achieve more by refusing to negotiate. The different sides should try to develop and cultivate attractive options to negotiating. For instance, they should be sufficiently satisfied with the status quo that they do not have to negotiate. If they have attractive options outside the negotiating process, they can win through stalling and delay. Being in a position to be more patient than the other side, they do not have to give in. Their negotiating stances becomes firm. They should not signal to the other side that they are willing to compromise. If they form coalitions, even with traditional enemies, it should be to break their opponents' will.

The problem with the realists' position is that, bound by incompatible positions, the sides have no room to compromise and stalemate is the inevitable result. Hardball tactics lead to intractability and preclude gains that all the sides involved in a conflict can win. Therefore, the advice that the mutual gains approach gives is to collaborate: Understand each other's views by working through hostility and mistrust by means of face-to-face meetings that narrow the range of disagreement, generate new options, and expand the potential benefits for all the parties. To achieve large gains, the parties must be willing to absorb small losses and sacrifice a portion of what they care about to achieve some greater goal.

Pragmatists question whether these methods really work. Long-standing adversaries will not easily come to agreement. Typically, their claims are based on fundamentally different views about who has endured greater harm, who has to make greater concessions, and whose needs are most pressing. The pragmatic perspective emphasizes the need to understand a dispute's history and content, as well as its social, political, and institutional context. The parties to a conflict are not just seeking an advance over the status quo but an advance proportionate to the weight and legitimacy of the long-standing claims they have. Any agreement reached has to satisfy an underlying sense of fairness. It has to overcome problems such as *reactive devaluation*, that is, the tendency on the part of antagonists to denigrate compromises offered by their opponents (Ross and Stillinger 1991; Kwon and Weingart 2000). Certainly, pragmatists encourage the parties to maximize joint value and expand the size of pie, but in the end, doing so does not eliminate the struggle to maximize each parties' own share, no matter how much bigger the pie has become.

[7]It can be argued that the differences between equity and responsiveness reside more in the minds of regulators, especially in the minds of agency attorneys, than exist in practice. Permitting at any facility is quite site specific, and alternatives to fixed rules such as arise in pollution prevention approaches are regularly available. In Aspen's two-track approach, every facility is free to achieve beyond-compliance status and participate in the more responsive track.

[8]Summary by D. Geffen of comments by participants and observations at Project XL Minnesota Lessons Learned for Moving Forward Meeting, held in Minneapolis, Minnesota, April 1, 1997.

[9]Letter from T. Zosel, manager of environmental initiatives at 3M, to S. Herman, assistant administrator, EPA Office of Enforcement and Compliance Assurance, expressing concern about reliance on "standard operating procedures" rather than "innovative thinking," Oct. 16, 1995.

[10]Hirsch (2001) argues that there are actually two types of XL projects—those that experiment with new regulatory approaches and those that try to fix specific instances where existing regulations break down. Most XL projects only try to fix specific instances where existing regulations break down. Very few are really experimental.

11

Creating a Platform for Experiments

We believe that experiments with alternative regulatory approaches that provide opportunities for learning should be carried out. On the basis of our analysis of the impediments to cooperative environmental agreements, we make a number of recommendations about how to carry out these experiments. Our proposals can be applied to Project XL or to regulatory reform programs that succeed it. They are broader in scope than XL and focus on reducing the transaction costs through better ways of dealing with superior environmental performance and ways in which the speed and efficiency of environmental accords can be negotiated. The U.S. Environmental Protection Agency (EPA) has addressed these issues as well, but we believe that its remedies fall short of what is necessary to revitalize Project XL or programs like it. Because our recommendations parallel the issues that EPA has addressed, we first turn to a review of EPA's reexamination of XL and the changes it has made.

Reinventing Project XL: Self-Criticism at EPA

EPA's XL managers, to their credit, recognized that they were having problems implementing Project XL. By August 1996, it was clear to them that the project at 3M Company's Hutchinson, Minnesota, plant was in trouble and that the difficulties 3M and other projects were experiencing could discourage companies from developing new XL proposals. EPA managers took a number of steps to remedy this situation. They redefined superior environmental performance (SEP) and tried again to explain its connection to flexibility. They also worked on the process the agency was using to make decisions. To this end, they held meetings and prepared and published a number of documents to provide greater clarity to project applicants. Our brief summary of their activities is not meant to be a complete accounting of their efforts, but we hope it is complete enough to highlight their main actions.

In September 1996, the EPA Office of Policy, Planning, and Evaluation (OPPE), which was responsible for administering Project XL, held a staff workshop, called XLpalooza, to discuss problems in XL and ways to solve them. In the same month, OPPE staff prepared a document titled *National Stakeholder Involvement in Project XL*, which stated that national environmental organizations had important roles to play in determining the direction of the program as a whole and developing individual projects (U.S. EPA 1996d). 3M withdrew its Hutchinson XL proposal in December 1996.

Two months later, EPA took the administration of Project XL and other reinvention projects from OPPE and established a new Office of Reinvention to coordinate and administer regulatory reform initiatives, including XL (U.S. EPA 1998a, 1998c, 1998f).[1] On April 1, 1997, EPA held a meeting in Minnesota to learn from the 3M–Hutchinson experience.[2] It published a mid-course correction in the *Federal Register* (62 FR 19872 or FRL-5811-7) on April 23, 1997, which clarified implementation guidelines for SEP, regulatory flexibility, and the stakeholder process, while soliciting more proposals.

Other important events centered on the efforts of state environmental control officials to encourage EPA to give the states more authority to move forward with reform and experimentation.[3] Later in 1997, EPA began a process to redesign and streamline the XL approval process, and it took steps to recruit new companies. The agency published another notice in the *Federal Register* on June 23, 1998, further clarifying the definition of regulatory flexibility and soliciting additional pilot projects (*Federal Register* 63 FR 34161 or FRL-6113-5).

As an aid to XL participants, EPA published several guides.[4] It released a guide for stakeholder involvement (U.S. EPA 1996g) in late 1996. It issued a best practices guide on its website to help sponsors develop XL proposals (U.S. EPA 1999a) on June 3, 1998. It trained its staff so that they could bring out important issues in XL pilots early (U.S. EPA 1998b). It taught them about mutual gains negotiating as developed by Fisher et al. (1991) in *Getting to Yes*. The agency's goal was to have decisions about XL pilots made in no more than six months.

Our list of EPA's improvement efforts is by no means complete, and the agency is still engaged in an ongoing process of continuous improvement.

Nevertheless, despite EPA's efforts to improve and accelerate decisionmaking, the transaction costs of individual projects remained high. According to Delmas and Mazurek (2001), it took on average 20.5 months to negotiate an agreement with business. The high transaction costs and delays were a result of "a lack of clarity in project guidelines—particularly those concerning superior environmental performance" and "poor coordination among federal, regional, and local regulators" (Blackman and Mazurek 1999, 5). The business community still regarded XL as cumbersome, bureaucratic, and prone to unnecessary delays. The number of XL project approvals did not appreciably increase, and participation by the business community continued to be disappointing (see Figure 10-1).

The XLpalooza Meeting

EPA's reexamination of XL started with a workshop called XLpalooza, which was held in New York City on September 11 and 12, 1996.[5] This meeting was highlighted by presentations from Fred Hansen, the deputy administrator, David Gardiner, assistant administrator for OPPE, and Jon Kessler, director of OPPE's Emerging Sectors and Strategies Division and head of regulatory reform efforts like XL. Hansen closed the two-day session with an overview in which he emphasized the importance of reinvention. He said that reinvention was critical because it was an effective tool in EPA's struggle against regulatory rollback. However, he admitted that EPA was concerned about the overall credibility of XL (see, for example, U.S. EPA 1996b, 1996h, 1997e). Many projects were languishing because of the length and difficulty of the approval process, and few additional facilities were showing a willingness to participate.

Hansen conceded that industry was "upset" and concluded that painful negotiations with companies like 3M (see U.S. EPA 1996a, 1996c) had made it difficult to attract additional applicants. Hansen put some of the blame for this state of affairs on the business community. He said that the reason XL often did not work was that industry was not as forthcoming or creative as it should be with ideas about SEP. Hansen also maintained that EPA had to rethink how it was dealing with national environmental organizations.[6] It was in the best interests of the program to involve them; however, the agency had not done a good job in defining their roles or involving them from the beginning.

During the XLpalooza meeting, Gardiner stated his vision for XL. He said that the initiative represented a laboratory in which EPA, state and local governments, and other stakeholders could experiment with alternative environmental strategies and produce tools and mechanisms from which future environmental protection strategies could evolve. He noted that XL was still in its formative stages, and he directed the attention of EPA's XL team to a number of key areas in need of improvement. The agency required better principles and approaches to SEP, more proactive stakeholder involvement, and a better process. By a better process he meant step-by-step implementation, improved cross-agency cooperation, headquarters leadership, better proposals, and a tracking system for making key decisions. He held that EPA had to arrive at its decisions more promptly, as a team, and with more accountability.

In his remarks, Kessler dealt with the decisionmaking process within EPA. He stressed the roles that the different elements within the agency should play. Management had to make policy decisions, staff had to raise issues, headquarters should guide projects in appropriate directions, and OPPE should ensure the integrity of the process. Kessler maintained that more effort should be put into working with project sponsors to improve and clarify proposals before submission. In what seemed like a reference to the 3M case, he held that the agency had to resist public relations pressures to come to hasty decisions and have numerous approved projects merely for the sake of having them approved. EPA should be willing to say "no" to projects at a very early date and to redirect them to other reinvention efforts when appropriate. Regarding the SEP that EPA should expect from individual pilots, Kessler made the point that XL was a very expensive program, and EPA had to assure that projects clearly yielded superior environmental results. Otherwise, XL was not worthwhile.

Individual EPA staff members raised a number of issues during the open discussions at XLpalooza. One noted that XL projects were supposed to be experiments and pilots, but EPA was not explicit enough about what the projects were testing and how to assess their success. Another pointed out that the uniqueness of the companies and their situations meant that it was not easy to transfer a specific project into a policy initiative. Still another said that without new laws EPA was handcuffed. It could not do macroprojects that had a chance to change the paradigm under which it worked. Finally, one EPA staff member said that XL's design flaw was that it required fitting projects into the "box" of business as usual and the existing legal structure, and that doing so was very difficult.[7]

The reassessment of XL that occurred at the XLpalooza meeting helped EPA set the direction for the corrective actions it took. It released the previously noted documents and guides (U.S. EPA 1998b, 1998d, 1998e, 1998g, 1998i, 1998j), restated its criteria for SEP, and tried to formally reengineer the process it used for making decisions. Assisting EPA in the reengineering effort were representatives from industry (Union Carbide and Dow Chemical), environmental nongovernmental organizations (Environmental Defense, Environmental Law Institute, and Citizens for a Clean Environment), and other concerned stakeholder groups. The changes EPA made are discussed next.

SEP and Flexibility

EPA tightened the logic of the quid pro quo, explicitly linking the flexibility it would grant to the amount of SEP a company offered.[8] Applicants had to demonstrate a "factual link" between SEP and flexibility: "the closer the factual link between the requested flexibility and anticipated environmental benefits," the more likely EPA was to approve a project (U.S. EPA 1997e). Rather than conduct experiments to explore the extent to which environmental benefits were a consequence of flexibility, the agency was looking for facilities that guaranteed successful outcomes from the start. These policies meant fewer risks, but reduced opportunities for learning.

The agency also redefined SEP. It separated it into two categories (*Federal Register* 62 FR 19872 or FRL-5811-7, April 23, 1997). The Tier 1 requirement was enforceable in the XL permit (e.g., subject to fines) and established a baseline to ensure that a facility's environmental performance was legally able to support a site-specific rule or some other type of rule. The baseline or "benchmark" for SEP was defined as environmental performance at least as good as it would have been had the facility operated under existing regulations or under future regulations, whichever was more stringent.[9] EPA assumed that voluntary controls on unregulated sources would remain in force and there would be "no credit for past good deeds."

Tier 2 was a nonenforceable voluntary commitment a facility made in the final project agreement (FPA) to achieve environmental performance superior to that established in Tier 1. Under Tier 2, a facility indicated what more it could do beyond what it had committed to accomplish under Tier 1. The *Federal Register* Notice (*Federal Register* 62 FR 19872 or FRL-5811-7, April 23, 1997) did not quantify how superior the Tier 2 commitments had to be, however. As evidence of superior performance, the facility could engage in pollution prevention, adopt other best practices, improve the quality of stakeholder involvement, or reduce compliance costs. Although the Tier 2 commitments in the FPA were nonenforceable, failure to meet all or most of them could result in termination of the agreement and the permit.

The approach to SEP that EPA took was an improvement over the one it replaced, but it still had serious shortcomings. The two-tier designation was an effort to clear up the uncertainty about requirements that were enforceable and belonged in the permit and requirements that were "voluntary" and belonged in the FPA. For a facility with a fairly stable level of production, a product mix that did not change much, few or no new product introductions, and no important new regulations on the horizon, the Tier 1 baseline became, in effect, its actual pollution releases. Only small and fairly unambiguous adjustments would have to be made for changes in output. For a facility like 3M–Hutchinson or Merck's Stonewall, the baseline determination would continue to be problematic because of production increases, changing product mixes, new product introductions, and new regulations on the horizon, such as maximum achievable control technology standards for emissions of hazardous air pollutants. In such cases, it would be difficult to define a workable unit of production.[10] Such issues would have to be resolved on a case-by-case basis with regulators exercising discretion. To company managers contemplating participation in an XL pilot, relying on a process this uncertain posed risks they might not be willing to bear.[11]

Improving the Process

EPA also tried to improve the process it used to make decisions. To eliminate roadblocks, which might result in delays, it set up teams to help project applicants prepare proposals and to screen out projects regarded as unsuitable as early as possible. To eliminate unnecessary transaction costs, it clarified the role of stakeholders. The March 1999 *Manual for EPA XL Project Teams* established timelines of between 200 and 345 days from the initial preproposal stage to the development of a permit and FPA. EPA offices reviewing a proposed pilot were to do so expeditiously, keeping these deadlines in mind.

To speed the flow of information and produce more punctual decisions, EPA modified its decisionmaking process. Yet the new process the agency fashioned still relied on a series of linear interactions and multiple iterations, with many offices and programs within EPA involved. Decisionmaking was consensus based, and therefore it had shortcomings for achieving negotiated outcomes (Coglianese 1997, 1999). The revised system continued to be complex. Documents flowed back and forth many times between the agency and an applicant in an iterative cycle, which compounded the possibilities for objections, modifications, delays, and gridlock. This process was not one likely to satisfy applicants like 3M, which were looking for a crisper management structure from EPA and a more rapid resolution of their cases. It was no substitute for a parallel team approach where all the parties—including state governments, local stakeholders, environmental organizations, and the federal agency—had the chance to simultaneously discuss issues in face-to-face encounters.[12]

Recommendations for SEP and Faster, More Efficient Decisionmaking

Though EPA dealt with the trade-off between SEP and flexibility and tried to improve the decisionmaking process, these issues did not go away. The average number of months it took to negotiate an agreement remained high, and the most important reasons continued to be an ambiguous standard for SEP and problems in how the agency made decisions. Our recommendations to improve the experimentation process mainly concern SEP and EPA decision-making, but we start with the legal issues.

Legal Authority

As our discussion of the legal issues has shown, there has been lively debate about the degree to which environmental laws are flexible enough to permit experiments in regulatory reform (Hirsch 1998). Enabling legislation would be useful to resolve these uncertainties. EPA could use authorization to grant variances from existing environmental laws and rules to pilot projects as part of an experimental program to test alternative approaches to existing regulation. There must be strong safeguards that environmental performance under the experiments would be no worse than under existing laws, but the legislation need not require SEP for each project.

Ideally, the enabling act should authorize funding to cover the costs of administering pilot projects and include funds to provide independent technical expertise to participating stakeholders when needed. If passed and implemented, such legislation would open the door to experiments that would test innovative ideas and approaches. It would eliminate a major roadblock to achieving cooperative solutions: conflicts over the legal authority to conduct specific pilots. The need to unleash greater creativity and innovation to tackle serious future environmental problems outweighs the risks that passing such legislation might weaken existing requirements.

The practical impediments to passing this legislation, though, remain high. Even in the absence of a new statute, EPA should use the discretion it has under existing laws to seek alternative solutions and approaches to dealing with SEP and to increasing decisionmaking speed and efficiency. We outline our proposals for dealing with these issues in the sections that follow. We do not wish to oversell our solutions because, as we have indicated, regardless of improvements, there are inherent difficulties to negotiating environmental accords. Nevertheless, we believe that our proposals, if adopted, would increase business participation in regulatory reinvention experiments and add to the learning. Only by seriously experimenting with new approaches is it possible to discover what is feasible and, possibly, to achieve substantial gains for the environment and the economy.

Better Ways to Deal with SEP

Confusion over the meaning of SEP has been a key barrier to implementing pilots. Under the quid pro quo incorporated into Project XL, before EPA could approve a proposed pilot, an applicant had to demonstrate that it was able to achieve SEP. How SEP was defined was not clear, however. For instance, was it in terms of pollution allowed or actual pollution emitted? Was it through pollution per unit of production, technological feasibility, or some other measure? Issues in defining SEP also included how to establish a baseline for improvement (should it be releases during a recent typical year or extrapolations into the future?), how to consider future expansion when determining superior performance, and whether past voluntary efforts would be rewarded. The definitions EPA used were not clear, they varied from case to case, and they remained ambiguous even after EPA redefined them.

Unless such issues are resolved, it will be difficult to reach agreements under Project XL or under programs that the federal government devises that are like it. There are conceptual problems in defining SEP and basic problems with the quid pro quo itself. How does EPA know if individual pilots have achieved environmental performance better than can be expected under conventional rules? Discretion has to be applied and much depends on the particulars of a facility's operations, the nature of its pollutants, and the environment into which these pollutants are released.

We propose two alternative approaches to dealing with SEP. The first is to drop the quid pro quo approach entirely, require only a level of environmental performance at least as good as regulations require (i.e., Tier 1), and emphasize the program's experimental nature, thereby removing a difficult-to-negotiate, contentious issue and making experiments more productive. The second is to create a somewhat more clear-cut, unambiguous definition of SEP. If standards defining SEP are better known and there is less uncertainty, an applicant will feel more secure in applying for a permit and EPA officials will have an easier time determining if the applicant is qualified. As in any negotiation, the likelihood of success will go up if the ground rules are clearly known in advance and if the objective or desired outcome is made as direct and unambiguous as possible.

In practice, however, such clarity may not be so easy to achieve. Indeed, both approaches are likely to face difficulties. The first collides headlong with EPA's institutional culture, which is to maximize environmental improvements whenever the opportunity arises.[13] The second poses obstacles for the business community, whose institutional culture makes it averse to accepting the risks of predetermined voluntary commitments without the potential for commensurate economic returns. By virtue of fixing the amount of SEP that has to be achieved, the latter alternative also reduces the amount of learning that can take place, and, as we show (see Appendix C), it is probably so complex that it would be hard to carry out.

Option 1: Hypothesis Testing and Learning—SEP an Outcome and Not a Requirement. We therefore start with the first approach. EPA can eliminate the requirement that a facility must guarantee a fixed amount of SEP beforehand. Doing so would remove a significant barrier to business participation.

Projects would be chosen based on the attractiveness of the regulatory innovations being tested and the extent to which they provided opportunities for learning. Projects would test hypotheses; for example, greater operating flexibility provided by facilitywide performance standards would over time result in better environmental outcomes and/or lower cost.

Not all tests of such propositions would be supported, however. The nature of an experiment is that the outcomes cannot be determined in advance. Furthermore, a particular idea needs to be validated by trying it a number of times, and every test will not necessarily show positive results. The kinds of tests carried out must be worth pursuing because of the broad benefits they promise to deliver—overall SEP in the long run, as the entire regulatory system is transformed and becomes more conducive to yielding higher levels of SEP, not just a few examples of the successful delivery of this result.

Safeguards would have to be built into this approach. Facilities should not have complete carte blanche to operate as they see fit. Permits would have to require that facilities' pollutant releases would be no greater than would be allowed if conventional environmental regulations were applied to all regulated sources. Guarantees would have to be in place that clearly spelled out the consequences for noncompliance. Facility managers would have to engage in good-faith efforts and accept the additional burdens that conducting experiments places on them.

Eliminating quid pro quo determinations of how much superior environmental performance should be exchanged for how much flexibility would encourage companies to come forward with new ideas.[14] It would change the nature of the innovations pursued from complex bargains to learning opportunities. Pilot projects would be viewed in experimental terms, and in accord with their experimental nature, the government would take a broad approach to encourage the use of performance-based standards, multimedia pollution prevention, innovative environmental management systems, stakeholder accountability, and other methods for achieving SEP and cost savings.

FPAs would have goals for achieving SEP. Though SEP would not be a precondition for project approval, there would be discussion of how and why goals for SEP had a reasonable chance of being attained. XL managers at EPA and state agencies would have the chance to choose the most promising projects. Some ideas may only yield environmental outcomes roughly equivalent to that achieved by existing regulation, but do so at measurably lower cost. The resources saved can be invested in other productive efforts, some of which can improve public health and the environment. Under Tier 2, SEP goals have to be met during the duration of a permit, but under our proposal, they would be in place for a longer period of time, which would encourage long-term investment.

EPA should carry out experiments at facilities whose sources have been grandfathered in.[15] Large numbers of these unregulated sources continue to operate throughout the United States long after Congress had expected them to be phased out. Facilitywide standards would be a way to tackle this problem. We believe that controls on grandfathered sources should not be introduced all at once. Permits should provide for a transition period so that plant managers are able to meet the new requirements in innovative, cost-effective ways.

The government, along with independent analysts, would have to monitor and carefully evaluate the experiments to determine if granting greater flexibility resulted in hoped-for improvements in the environment and the economy. A proper evaluation of these experiments is by no means an easy task. It would be necessary to make an adequate determination of pollutant releases at a facility during the course of the experiment and before its initiation and to establish a baseline. It would be necessary to compare releases during the course of the project with baseline amounts and with SEP goals in the FPA. There would be problems encountered in scaling baseline pollutant releases to changes in production levels and product mixes. Analysis of the environmental benefits of a particular alternative regulatory approach would require an estimate of how the facility would have performed under otherwise applicable regulations.

When evaluating the outcomes of experiments, the underlying problems inherent in any determination of SEP exist.[16] Evaluations are likely to be subject to interpretation. They may result in controversy. Many pilots will have to be carried out under different circumstances to deal with this problem—that is, to determine if a particular approach really does work.

Although we believe that the change in SEP we are proposing would advance goals to which EPA, environmental advocacy organizations, and businesses are committed, the institutional issues discussed previously (see Chapter 10) pose problems that make it unlikely that this approach will be adopted. EPA seems to be particularly averse to the risks that individual reinvention projects will not exhibit clear, tangible amounts of SEP that are guaranteed in advance. Therefore, as an alternative to this approach, we consider a less venturesome and potentially less rewarding option that we believe still would permit learning, but less so than in the first instance.

Option 2: Establish Clear Rules for Assured Environmental Benefits. If the institutional barriers to implementing a hypothesis-testing and learning approach are too great, EPA can ask for clear, up-front guarantees of SEP for all pilot projects but eliminate an important source of uncertainty for businesses by providing more definite guidelines about what SEP means. The guidelines for SEP would not be linked to the regulatory flexibility sought by an applicant. Projects would be chosen based on the attractiveness of the innovations being tested, and bargaining over SEP for flexibility would be reduced or eliminated.[17] On the basis of an analysis of the cases in this book, we make suggestions about how EPA might arrive at a less ambiguous definition of SEP (Appendix C), the details being less important than the approach, because as experience accumulates on the various XL projects and more analysis is done, the specifics would have to be modified. The purpose is to draw lessons from existing pilots that would be useful in helping EPA move forward.

As can be seen from the example in Appendix C, establishing a more explicit definition of SEP would be difficult. Regulating pollution is a complex, site-specific challenge in which one size does not fit all. There are likely to be exceptions and the ensuing bargaining process would entail high transaction costs. Nevertheless, with fixed quantitative guidelines for SEP, negotiators are less likely to become bogged down in controversy over how much SEP is needed to pay for the operating advantages a company is seeking. Companies will know in advance what is expected of them, however stringent and restrictive they might consider these requirements to be.

If EPA is not willing to adopt the more venturesome, hypothesis-testing and learning option (the first proposal we made), this alternative (clear rules for assured benefits) might be acceptable. However, providing explicit guidelines for SEP does not solve all problems. Facilities seeking an alternative permit still will have to be certain that they can meet the enforceable limits this approach requires. Many companies will regard having to meet such limits as risky so that explicit definitions of SEP actually will prevent them from participating.[18] With restrictive definitions of SEP in place, applications for experimental permits actually might go down. The potential benefits that companies can achieve might not justify the cost of taking on "stretch" goals and the risks entailed in not achieving them.[19] Rather than exploring innovative environmental management strategies that will test new ideas that might not work, companies will focus instead on less pioneering and risky notions. Though some learning still will take place, less is likely than under the first option where hypothesis testing is central and SEP not guaranteed.

Speed and Efficiency in Decisionmaking

Eliminating the quid pro quo or providing more specificity in how SEP is defined would help revive a program like XL. The pressure on negotiators would be reduced, but there still would be issues of judgement and questions about how this judgement would be exercised. In instances where EPA officials are in a position to exercise discretion, questions would remain about how they would do so. Therefore, it is as important to address the process questions of *how decisions will be made* as it is to address the substantive issue of the framework for SEP and flexibility that will be used.

So far, the transaction costs for parties participating in Project XL have been unacceptably high. Although EPA has established—more than once—ambitious timetables for the development and approval of individual pilot projects, these timetables will not be met unless there are changes in the way the agency makes decisions. Globalization, changes in technology, and increases in environmental impact make timely decisionmaking an imperative for businesses. The large commitment of time required of other stakeholders that participate is a burden for them as well. EPA should nurture a culture where both the speed and the quality of the decisions it makes are important (Kessler and Chakrabarthi 1996). To build interest in negotiated environmental agreements not only among business firms but among environmentalists, state agency staff, and members of local communities, EPA should aggressively strive to meet the schedules it has established for reaching environmental agreements (Guth and MacMillan 1986).

A considerably faster decisionmaking process would remove a disincentive for companies to become involved in XL or programs like it. If businesses, governments, and other groups in society are to work together to solve common environmental problems, they have to employ new tools for prompt and effective decisionmaking (Bower and Hout 1988; Dumaine 1989; Cushman and King 1991). Toward this end, we urge EPA to adopt the following changes in its decisionmaking.

The Role of EPA Headquarters. XL staff members in the various program and other offices at EPA headquarters should stop trying to micromanage individual pilot projects. The expertise of headquarters should be involved only in very exceptional cases where there is serious, overlooked risk to the environment or exceedingly important legal issues to resolve. The role headquarters plays is a critical one, but it is not to approve individual pilots. Rather, it is to structure the experiments to be carried out by deciding on the innovative ideas the agency wants to test and on the number of experiments it needs to conduct each test. Headquarters should then assign the resources and responsibility for implementing these pilots to EPA regional offices.

Because headquarters has responsibility for setting policy, it must carefully evaluate the performance of pilots to determine if they have been successful and how they fit into a larger concept of regulatory reform. After the fact, working with national environmental groups, businesses, and other stakeholders—perhaps by creating and empowering advisory groups—headquarters should interpret and draw significance from the results. In this way, it should derive practical lessons about how to bring a new approach to environmental management into being, one that is both better for the environment and better for the economy.

Selecting Experiments. Because the role of EPA headquarters is to structure the experiments so that learning takes place, it is necessary to consider the type of experiments that best facilitate learning (Dougherty and Pflatzgraff 1970; Arthur 1997; Foray and Grubler 1996; Freeman 1996; Gould 1991, 1996; Ruttan 1996; Tushman and Anderson 1997). At one extreme is the idea that learning is not possible without a large number of experiments and without preselection of projects. There is too little information and too much uncertainty to be very careful about choosing which experiments to carry out in advance (Tushman and Anderson 1997). At the other extreme is the idea that only the very best experiments should be pursued. Action should not be taken without appropriate foresight and initial screening that reduces unwanted surprises.

The Clinton administration's March 1995 plan to reinvent environmental regulation led EPA to probe in many categories *but to be very selective about pursuing projects in any particular category*. The plan to reinvent regulation had 25 "high-priority" action items divided into 18 "improvements to the current system" and seven "building blocks" for a new one. Under the reinvention mandate, EPA was able to expand the original list into more than 60 programs (U.S. EPA 1996b, 1997c, 1997d,) and claim that no list could adequately represent all the reinvention activity in which it was engaged. With so many programs under the reinvention umbrella, integration was a challenge (U.S. EPA 1997d, 1998a, 1998c, 1998f, 1999b). The highest-profile programs were the Common Sense Initiative and Project XL. The Common Sense Initiative's focus was on industry sectors, but during its first two years (1994–1996) only two projects were approved (Davies and Mazurek 1996). XL's focus was on permitting individual facilities, and although it was supposed to produce 50 agreements a year, it did not achieve this goal until 2000, five years after it was initiated.[20]

More learning would take place if EPA probed in fewer categories but carried out more and better experiments in each. EPA also should better integrate the changes it is making. It should look for the linkages between programs. With more focus on fewer programs and increased integration of these programs, the overall thrust of the agency's efforts is less likely to be dissipated.

Delegating Authority. EPA should emphasize the experimental nature of projects, while downplaying their precedent-setting character (Kessler and Chakrabarthi 1996; Gersick 1991; Eisenhardt and Tabrizi 1995). Individual pilot projects are experiments. Only as a group are they likely to give rise to changes in national policy. To bring about more rapid decisionmaking, the regions and not headquarters should have the right to approve individual pilots and manage them.[21]

In negotiating environmental accords, broad-based stakeholder teams from different institutions and organizations should provide input to the regions (Kessler and Chakrabarthi 1996; Eisenhardt and Tabrizi 1995).[22] Intel, Merck, and Weyerhaeuser had a more inclusive process than 3M, and most of the literature on cooperative agreements seems to agree that more-inclusive processes work better (Long and Arnold 1995; Bingham 1986). Thus, in designing regulatory pilots, the federal government should continue to rely on stakeholder teams that consist of individuals from local communities, state pollution control agencies, regulated facilities, relevant EPA regional offices, and, if needed, experts from headquarters. Because professional facilitators can play useful roles in drawing these parties together, EPA should make funds available for them. If there is agreement among those who have to play a role in carrying out a pilot, it will be better executed. Though stakeholder teams should be intimately involved, EPA's regional offices would be the *final* decisionmakers.[23]

Parallel Processes. EPA should open the capabilities for innovative problem solving within the stakeholder teams so they are better able to help the agency design pilots. To enhance communication and coordination, the teams should meet together often and as a group, preferably face to face. The teams should minimize the serial passing of documents from one group to another because delay is likely and misunderstandings are almost certain when this type of exchange takes place. Instead, they should maximize the use of parallel processes in which people from different groups are given the chance to work together collegially.[24]

In time-pressured situations where there is uncertainty and risk, members of teams need the ability to quickly analyze and communicate information so that issues under their consideration move rapidly toward resolution (Fidler and Johnson 1984; Grin and Van Den Graaf 1996). They need to coordinate diverse concerns and pool knowledge to be flexible and responsive. They should rely on overlapping, concurrent, and parallel processes so that the opportunities for delay and deadlock are limited. Overlapping stages spread real-time information from person to person. They increase flexibility and provide for iterative thinking, testing, and probing. Frequent, close interaction yields repeated evaluations and assessments of solutions' strengths and weaknesses, and thus it fosters rapid reconceptualization of ideas. Conflicts are likely to be reduced, and misunderstandings less likely to persist.

Business Involvement. In accord with the experimental nature of pilots, businesses then can test the use of performance-based standards, multimedia pollution prevention techniques, innovative environmental management systems, stakeholder accountability, and other ways of achieving significant cost savings and better environmental results. They should assign permanent staff with the requisite capabilities to design and carry out these experiments. Some of them are likely to be lengthy, and the commitment of company staff must be for the long term. For businesses to commit staff for the long term, they must see concrete economic rewards and improvements in the regulatory system.

Businesses also need a greater appreciation for the government's point of view. Governments can strive to make decisionmaking more rapid, but governments cannot proceed by abrogating legitimate political constraints. In being cognizant of the wisdom built into the system, businesses must tolerate less than optimal efficiency out of a respect for due process and checks and balances.

Businesses that expect to negotiate environmental agreements also cannot withhold vital information if they expect to take advantage of advanced permitting opportunities. They have to come forward with information promptly if they expect government to make decisions in an expeditious manner. Thus, they will need to disclose nonproprietary information that they might otherwise consider sensitive.

Conclusion

The recommendations we make in this chapter would not be easy to implement, but we believe they are worth pursuing. Changes such as these will go a long way toward ensuring that future attempts to experiment with innovative approaches to environmental management will be illuminating, productive, and successful. EPA, bound by its institutional history, may prefer the more certain, if less rewarding, path of tangible, guaranteed environmental improvement. Businesses may be reluctant to experiment if the transaction costs remain high and improvements are not made in how EPA makes decisions. But both parties have to seriously consider moving in a more exploratory direction. Though it may be risky, it is likely to yield substantially greater benefits for the environment and the economy.

Notes

[1] In 1999, the Office of Reinvention combined with the Office of Policy to form the Office of Policy and Reinvention.

[2] Summary by D. Geffen of comments by participants and observations at Project XL Minnesota Lessons Learned for Moving Forward Meeting, Minneapolis, Minnesota, April 1, 1997.

[3]ECOS 1997a, 1997b, 1997c, 1997d; letter from ECOS to C. Browner, administrator of EPA, and F. Hansen, deputy administrator of EPA, expressing shock at EPA's unilateral decision to immediately pull back from review the draft agreement to pursue regulatory innovation, February 26, 1997.

[4]For a look at several of the EPA documents, go to "XL improvements" at http://www.epa.gov/projectxl.

[5]The discussion that follows is based on EPA's meeting summary, minutes, and overheads (including "Superior Environmental Performance—Briefing," "XL Project Development Process—Briefing," and "Project XL Tracking System Description") from the XLpalooza meeting, College of Insurance, New York, September 11–12, 1996.

[6]U.S. EPA 1996d; overheads, "Involvement of National Environmental NGOs in Project XL," presented at the XLpalooza meeting cited in Note 5.

[7]Meeting summary, minutes, and overheads presented at the XLpalooza meeting cited in Note 5.

[8]See *Federal Register* Notice 62: 19872, April 23, 1997. EPA revisited this issue in June 1998 (*Federal Register* Notice 63: FR 34161 or FRL-6113-5). This *Federal Register* Notice reiterated EPA's preference for linkage: "Whenever a project also meets the other applicable facility or community decision criteria, EPA will aggressively offer the necessary flexibility to produce superior environmental performance and promote greater accountability to stakeholders."

[9]Scaled to take into account production changes, Tier 1 was calculated as pollutant releases per unit of production, but if a foreseeable tightening of applicable regulations were likely, it would be based on the new standards, which were not yet in force. Pollution prevention and inputs to the production of toxic materials were not excluded from Tier 1.

[10]There were other potential problems. For example, what if an XL-permitted facility phases out one product whose production is particularly amenable to a very highly efficient control system and replaces it with a product with production processes that yield pollutant releases more difficult to control but *less* toxic? Most industries have more stringent regulations in the pipeline (for example, the maximum achievable control technology rules for hazardous air pollutants). When, and how, would these rules be phased in? Facilities with substantial voluntary controls on unregulated emissions units under Tier 1 would have to accept full regulation of these units or find compensating reductions elsewhere. Moreover, how would EPA treat a situation where the facility reached the baseline by means of a high level of process and control efficiencies, which were not sustainable in the long run? Additional uncertainty came from the Tier 2 requirement. Potential applicants had to ask themselves how much better than the baseline they would have to perform to gain approval.

[11]The benchmark to which 3M wanted to adhere was simpler to administer—environmental performance better than existing regulations required, adjusted as regulations and their applicability changed.

[12]Carol Browner (1998), former administrator of the agency, maintained that EPA was learning how to do XL projects and that more of these projects would be successfully completed over time. EPA was trying to balance its role of controlling industry behavior and enforcing regulations with a new role of forming partnerships with industry. In the early XL cases, EPA demonstrated that it would require a high level of environmental performance in exchange for greater operating flexibility. It was put-

ting the burden of proof on an applicant to convince the agency that a proposed pilot could achieve performance superior to what otherwise could have been achieved under existing requirements.

[13]Spence and Gopalakrishnan, who analyzed 36 industry-sponsored XL proposals, 24 of which had been rejected or withdrawn, found that the proposals that offered more tangible and guaranteed environmental benefits in the form of reduced pollution emissions were the ones most likely to be accepted. Satisfaction of the SEP criterion with guaranteed pollution reduction was the surest path to having an accepted XL proposal (Spence and Gopalakrishnan 1999, 34). Even with unanimous stakeholder support, if pollution reduction was not guaranteed, EPA was unlikely to approve a project.

[14]EPA's reform efforts should foster broad benefits rather than narrow ones from a few site-specific, quid pro quo pilots, but there still might be a role for the latter type of permit when the flexibility granted provides the incentives to make environmental improvements that otherwise would not be made (Merck, for example).

[15]3M's grandfathered and hence unregulated sources at Hutchinson created significant complications in negotiating the XL permit, especially because of the voluntary controls 3M had put in place.

[16]The 3M–Hutchinson XL proposal tried to take this problem into account by requiring 3M engineers to estimate, for each project or modification at the plant with an environmental impact, what fraction of the environmental performance benefits that occurred could be attributed to the XL permit compared with what could have been achieved under a conventional permit.

[17]These are benefits that this option shares with the first one.

[18]Prior to entering XL, Weyerhaeuser was confident it could meet its effluent reduction with well-understood process improvements already in the preliminary stages of implementation, and Merck knew that its boiler conversion would reduce criteria pollutants so greatly that it would have no problems, but 3M was not comfortable making commitments that it might be unable to meet. Hutchinson was making many changes, and management was not entirely sure what would happen.

[19]The greater the SEP a company proposed, the more resources it would have to invest and the higher the risk that it could not achieve the SEP it promised.

[20]By the end of 2001, EPA had approved 51 XL pilot projects, 49 of which were still in force. Of these 49 projects, 14 were with local or regional governments, four were with state governments, five were with federal entities (four with Department of Defense and one with the National Aeronautics and Space Administration), one was a university coalition, one was a farm cooperative (eggs), 23 were industry projects, and one was a mixed project involving research laboratories in government, universities, and industry. The two terminated pilots were with the Jack M. Berry Corporation and the HADCO Corporation. In both cases, management of the facility changed hands. Cargill, which had taken over operation of the Berry facility in 1997, could not reach agreement with EPA and Florida about continuing the project, and the two agencies terminated the pilot in 1999. HADCO withdrew from Project XL more recently after being acquired by Sanmina/SCI, which reassessed the project after HADCO had encountered problems implementing it.

[21]When negotiating teams for Intel, Merck, and Weyerhaeuser had hammered out an agreement with EPA's regional offices, they thought they were close to achieving a

final agreement. Instead, they experienced significant delays while the project had to be renegotiated with other groups at EPA headquarters.

[22]Our proposals for improving the process of decisionmaking in XL or a program like it are in the spirit of what Weber calls "pluralism by the rules." He writes: "Instead of consultation and negotiation to discover the lowest common denominator around which everyone can unite, and instead of constructing a maximum winning coalition, the by-the-rules framework focuses on maximizing participants' preferences and practicing inclusion. Rather than negotiation marked by information deficits and limited consideration of policy alternatives, there is shared information, a concerted search for more reliable information, and a more robust search for policy innovations and alternatives" (Weber 1998, 17).

[23]The stakeholder teams with which the regions would work would function as "decision aides" not as decisionmakers (Gregory et al. 2001).

[24]Delmas and Mazurek (2001) found that the typical project had five different stages, each entailing considerable costs: (1) preparing a preliminary proposal, (2) conducting stakeholder negotiations, (3) interacting with local regulators, (4) interacting with the state government, and (5) obtaining final EPA approval. If these five stages were done more in parallel, EPA's decisionmaking would be improved and the time per project would go down. Of course, EPA would have to provide the regional offices with the resources they needed to carry out this approach.

Appendix A

The Comparable Actions Test

The following material is reproduced from the website of the U.S. Environmental Protection Agency (EPA): http://www.epa.gov/projectxl/3mhut/121396.htm (accessed May 22, 2002). It is a reproduction of EPA's definition and elaboration of the Comparable Actions Test (CAT). The CAT was EPA's counterproposal to the 3M Company and the Minnesota Pollution Control Agency (MPCA). It would have been the basis for the Project XL permit that EPA was prepared to give 3M at Hutchinson, Minnesota.

Comparable Actions Analysis

The permittee and the MPCA shall complete an air quality regulatory analysis report of a comparison between the actual emissions of each pollutant listed in Section II.A.1, Tables 1 and 3, and the Comparable Action Emissions—what those actual emissions would have been if comparable actions would have occurred under all of the existing requirements of the Clean Air Act. Any activity that would have required PSD (prevention of significant deterioration) or New Source Review, would require action because of new regulations, or which resulted in a voluntary reduction in emissions shall be specifically delineated in the analysis.... The regulatory analysis report shall be completed on April 1, 1999; April 1, 2000; April 1, 2001, and in conjunction with the reissuance of this permit. If actual emissions from the facility in tons per year are greater than emissions under the comparable actions test, the permittee shall do the following:

1. Within 60 days of the report, explain to the MPCA why the facility's actual emissions are greater than emissions under the comparable actions test, and report the exceedence and explanation to the permittee's stakeholders and EPA no later than the next scheduled stakeholder meeting; and

2. If actual emissions are greater than emissions under the comparable actions test by less than a de minimis amount or the exceedence is the result of startup, shutdown or the emergency. "Emergency" means any situation arising from sudden and reasonably unforeseeable events beyond the control of the permittee, including an act of God, that unavoidably increases emissions and that requires immediate corrective action to restore normal operation at the facility; or

3. If actual emissions are greater than emissions under the comparable actions test by more than a de minimis amount or were not the result of a startup, shutdown or an emergency as defined in Part 2, the permittee shall, within 90 days: (a) submit a plan to the MPCA to reduce actual emissions from the facility in an amount to offset, during a period not to exceed twelve months from plan approval, the emissions that were greater than emissions under the comparable actions test. Upon approval by the MPCA, the permittee shall implement the plan; or (b) if the permittee must install control equipment or undertake other pollution prevention or reduction activities to prevent future emissions greater than emissions under the comparable actions test, the permittee shall submit a plan to the MPCA which demonstrates a commitment to reduce actual emissions from the facility to or below the level of emissions under the comparable actions test during a period not to exceed twelve months from plan approval. Upon approval by the MPCA, the permittee shall implement the plan.

4. If the permittee fails to meet any requirements of Part 1 or 3, or fails to substantially implement an approved plan under Part 3, the permittee is in "significant noncompliance" with the Minnesota–XL permit within the meaning of Minn. Stat. Section 114C.12, subd 5(1). The MPCA may amend the permit, revoke the permit, or enforce the permit in any manner provided by law.

Comparable Actions

For each activity after 9/1/96 which would have, under the CAA [Clean Air Act], required PSD or NSR [New Source Review] review, been subjected to newly promulgated regulations, produced a voluntary reduction in emissions or resulted in lower emissions than required under the CAA, the Comparable Actions Analysis will include an addition or subtraction from the actual facility emissions according to the following table. (In all cases a difference between two emission levels is numerically an absolute value.)

New Emission Sources

Emissions greater than regulatory limit emissions. Subtract the difference between the actual emissions and the emissions that would occur if comparable production occurred from an emission source operating at the regulatory limitation.

Emissions below the regulatory limit emissions. Add the difference between the actual emissions and the emissions that would occur if comparable production occurred from an emission source operating at the regulatory limitation.

Modified Emission Sources

Emissions greater than the emissions resulting from the regulatory limits imposed by the modification. Subtract the difference between the actual emissions at the time of the modification and the emissions that would occur at the regulatory limitation and comparable production levels.

Emissions less than the emissions resulting from the regulatory limitation imposed by the modification. Add the difference between the actual emissions and the actual emissions after the modification at comparable production levels.

New Regulations Applicable to Existing Sources

Emissions greater than the emissions resulting from the regulatory limitation imposed by the new regulation. Subtract the difference between the actual emissions and the emission that would be allowed by the new regulation at actual production levels.

Emissions less than the emissions resulting from the regulatory limitations imposed by the new regulation. No action unless a physical or operational change was implemented to reduce emissions, then add the difference between actual emissions and the new regulatory limit at comparable production levels.

Voluntary Reductions in Emissions from Existing Sources

Add the difference between the actual emissions before the voluntary reduction and the actual emissions after the voluntary reduction at comparable production levels. Reductions in emissions at one unit which result in an equal increase in emissions at another unit are not considered voluntary reductions.

Netting or Offset Calculation

For any actions which trigger a regulatory requirement other than NSPS [New Source Performance Standards] or MACT [maximum achievable control technology], the use of netting or the use of Emission Reduction Credits which are allowed under the Clean Air Act will be allowed under the "regulatory limitations" of the Comparable Action Analysis.

Actual Emissions. The actual emissions that occurred from the facility during the calendar year prior to the Comparable Action Analysis.

Comparable Action Emissions. The emissions that would have occurred if the facility operated at comparable production levels under the individual unit specific requirements of the Clean Air Act and voluntary reductions that occur after 9/1/96 were not made.

Regulatory Limitations. These are the unit specific regulations that would apply to the facility under the Clean Air Act at the time an analysis is conducted.

Actual Production Levels. These are the actual production levels that occurred from the specific production unit during the period of the analysis.

Comparable Production Levels. This is a level of production that for unchanged products or technology is the Actual Production Level and for new products or technologies is that which would be comparable to producing that same product with generally representative industry practices.

Magnetic Tape Averaging System
(to be placed in Part IV.B of the permit)

The standards of Subpart EE of 40 CFR 63 for magnetic tape operations are varied by issuance of this permit. However, the permittee shall maintain its total actual HAP [hazardous air pollutant] emissions from magnetic tape operations to less than or equal to 90% of the total allowable HAP emissions under Subpart EE. Verification of compliance with this provision will be determined from an annual Magnetic Tape MACT Analysis in which the total actual emissions of HAP from magnetic tape operations in the previous year will be compared to 90% of the total allowable HAP emissions under Subpart EE at the actual production levels in the previous year. In completing this analysis, the permittee may group emission units and may use any compliance option available under 63.703(c) in determining its allowable emissions. The Magnetic Tape MACT analysis is due by April 1 of each year, beginning April 1, 1999. The permittee shall maintain records necessary to conduct this analysis. In the event that an existing magnetic tape emission unit is converted to the manufacture of a non-magnetic tape product, that emission unit will be considered a new emission source under the Comparable Actions Test.

RCRA Subpart BB Language

Amend Section II.B.4 of the permit to read as follows:

The air emission standards in Minn. R. 7045.0648 (air emission standards for equipment leaks) are varied by issuance of this permit. The permittee will provide five (5) tons of VOC emission offsets each year as part of its comparable actions test under Section IV.B of this permit until such time as the permittee demonstrates and the MPCA approves, using the procedure under Section III.C.1 of this permit, a different amount of offsets.

Appendix B

Next Steps at 3M–Hutchinson

When Project XL ended, 3M Company's Hutchinson, Minnesota, facility went through a number of changes. It was vying with other 3M production facilities on a variety of dimensions (for example, safety, quality, service, and cost) to be the site of choice to make adhesive tape products. It was capable of using many technologies and making products for a variety of markets—from health care to industrial, consumer, transportation, safety, and chemical. In 1998, 3M–Hutchinson had 1,750 employees. This number was down from a peak of more than 2,000. Aside from normal attrition, nearly 300 employees took early retirement and there were 150 new hires. Hutchinson's reputation within the company remained strong. It was making transparent tapes, microabrasives, and surface mounts (which held computer components together) for diverse customers in industries such as pharmaceuticals, fiber optics, paper mills, and utilities throughout the world.

Its goals for 2000 included the establishment of a new wing in the North Plant part of the facility. The new wing would be a quarter of a mile long. It would have a new coater that would be able to make numerous products for a variety of 3M divisions. In 1998, the number of active coaters in the North Plant had been reduced from six to four, but the coaters continued to be very

This appendix is based on notes from a briefing given to the Pilot Project Committee by a 3M Company staff member to update the committee on progress at Hutchinson. The briefing took place on May 12, 1998.

busy, running 24 hours a day and producing as much volume as the six audiovisual (AV) coaters used to make. Two of the coaters continued to make AV tapes for the reproduction of movies. Another coater was making a pharmaceutical product. The fourth coater had the capability to make the backing for the lithium polymer batteries that 3M was developing for low-polluting vehicles. 3M still was planning to completely exit from magnetic tape production because it remained convinced that digital technologies ultimately would replace AV tape. At the North Plant, its plans continued to be the same. It would continue to move into new high-performance, money-making adhesive tape products.

Since the collapse of Project XL, 3M had amended its permit at Hutchinson 10 times to accommodate six equipment changes. From 1995 to 1998, four engineers and one supervisor had spent approximately half their time on permitting, and the company had paid consultants about $130,000 in permitting related assistance. If production was measured as 3M previously had measured it in terms of square yards, it peaked in 1996, having turned down in 1997. However, if measured in terms of the revenues generated or the amount of pounds packaged, production continued to rise in 1997.[1] In 1996, magnetic tape production was at a peak, and the solvent-recovery system was heavily in use. The tonnage of volatile organic compounds (VOCs) emitted slightly increased, just as 3M managers had warned, but fell to all-time lows in 1997, as they promised (see Figure B-1).

The facility was adding high-performance tape products that commanded higher monetary return in the market. Along with the introduction of these

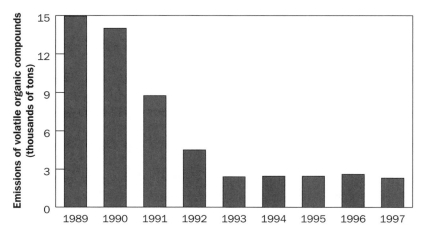

Figure B-1. Emissions of Volatile Organic Compounds at 3M–Hutchinson

Source: 3M Company.

new products, it also was trying to eliminate the use of solvents entirely. New products were mostly low in emitting VOCs and, because they replaced AV production, which was hazardous air pollutant intensive, they reduced overall harm over the product life cycle. To accomplish its pollution reduction goals, 3M was using hot-melt technology where feasible. It also increased the number of water-based coaters at the South Plant from one to three. Pollution prevention efforts continued as they both contained the costs of raw materials and reduced waste.

Nonetheless, 3M environmental staff members still felt that an XL permit would have helped. 3M was operating the Hutchinson facility in a mode of extreme sensitivity to market conditions and very rapid product change. Given the amount of change involved, the company would have been better off if it could have managed its time and people better. Although working with the Minnesota Pollution Control Agency on Project XL facilitated the permitting that took place at Hutchinson after the XL deadlock, the permitting process continued to be costly to the company and counterproductive as well, because plant engineers were diverted from more productive work by the need to meet 3M's compliance and permitting obligations. In addition, 3M staff members believed that they missed opportunities because new source review under the Clean Air Act forced them to focus on one source at a time, rather than to view the plant and its environment holistically and to consider long-term gains that they could achieve for the environment and their company. They could not take into account interconnections between sources and manage them from a systems perspective, one which they continued to believe was promising and had potential.

Note

[1]This apparent anomaly is because, when measured by area, magnetic tape products are made in much greater quantities than adhesive tape products but also have less value added and sell at a much lower price per unit of area. Consequently, the gradual phasing out of AV tape production would result in lower square yards of tape produced. This example illustrates the problems associated with defining a proper unit of production to measure pollution releases per unit of production.

Appendix C

A More Unambiguous Definition of Superior Environmental Performance?

We explore here how the U.S. Environmental Protection Agency (EPA) might introduce a clear rule for superior environmental performance (SEP) in order to reduce an important substantive barrier to developing regulatory reinvention experiments. This problem is a difficult one, and the ideas sketched below are not meant to be a definitive solution to this problem but only an example of how such a solution might take shape. EPA could use a common law–like approach wherein it would rely on decisionmaking in existing Project XL projects as precedents. The precedents established in earlier projects would be used as guidelines for determining whether applicants are meeting SEP thresholds. There would be full understanding that, as additional projects are considered and new precedents established, the thresholds would evolve; they are not set forever but would change in the organic way that law transforms itself in a common law system. In the example developed below, the early cases from Project XL are used as precedents to establish thresholds for SEP. These are safe harbors, which provide facilities capable of meeting them fast-track negotiating. Facilities willing to guarantee lesser amounts of SEP still could propose XL projects but they would have more hurdles to get over. This plan for SEP, derived from the Intel Corporation, Merck & Company, 3M Company, and Weyerhaeuser Company cases, has eight parts.

We thank Dan Farber and Dennis Hirsch for their reviews of the ideas presented here and for their suggestions for improving them. One of the authors of this book, A. Marcus, created this example.

1. Reserved for High Achievers

Project XL and other programs like it would be reserved for high achievers. There would be a bar, which facilities have to climb over, to be admitted. With a yardstick for admission, companies would have an incentive to make improvements in order to qualify. The Massachusetts StarTrack program uses this principle, and EPA is using it—quite successfully—in the Performance Track program (Nash 2001).

The facilities allowed to participate should have a record that is better than mere compliance with existing regulations. A number might be published in a *Federal Register* Notice that defines the meaning of this record. For instance, on the basis of the 3M, Merck, and Weyerhaeuser cases, proof might be that current pollutant releases are at least 25% below allowed releases in at least one medium (e.g., air pollution), and perhaps more than 35% below allowed releases if the percentages of lower releases from more than one media (air, water, and land) are combined.

This test is one that 3M, Merck, and Weyerhaeuser easily passed. If the facility is a start-up, as in the Intel case, and has no current actual emissions, then the company's other facilities—most similar in nature to the one in which it is seeking an XL permit—should be scrutinized to see if they meet this criterion.

The numerical value for this admission bar should be the most flexible of all the criteria we propose. EPA should select companies with proactive, enlightened environmental management practices, and simple quantitative standards provide only a rough selection screen. If a company cannot meet this criterion, it should be able to petition EPA and gain admission as a special case.

2. Lower Than Allowed Emissions

After meeting the requirements for admission to the program, a facility's XL permit and final project agreement application would be judged on the basis of a number of SEP criteria. The first is the difference between the pollution limits in the facility's proposed XL permit and pollution releases currently allowed by its existing permit. Again, an exact amount might be published in a *Federal Register* Notice and used for this determination. Businesses seeking greater certainty would not be subject to EPA discretion on a case-by-case basis.

The Intel, Merck, 3M, and Weyerhaeuser precedents suggest that emissions limits in an XL permit should be at least 40% more stringent than current allowable emissions in one environmental medium or, as an alternative, perhaps more than 50% more stringent if emissions from more than one medium are combined (e.g., 25% lower air emissions and 25% lower water discharges). This test is one that Merck, 3M, and Weyerhaeuser met (3M because so many of its sources were grandfathered in). If there are special circumstances, an applicant could petition EPA for an exception, with the burden of proof being on the applicant to demonstrate why the exception is needed.

3. Limits on Actual Pollution

Permit limits do not capture actual pollution, however. Thus, there should be another SEP test: reductions in either actual pollution or in pollution per unit of production. In making this determination, a distinction should be drawn between pollutants that are so toxic and hazardous that no increase in their level can be tolerated and pollutants that law and regulation allow to grow. Another distinction might be drawn, although it would be hard to draw it other than in air pollution cases where the law already makes this distinction in a straightforward way; this distinction would be between facilities in sensitive areas (nonattainment areas or areas adjacent to national parks and forests) and nonsensitive areas (most attainment areas) where these conditions do not apply. As long as the pollutants involved are not extremely hazardous or toxic, a facility in a nonsensitive area has and should continue to have the right to increase its pollution to accommodate growth in production.

The 3M and Weyerhaeuser cases fit this category. Their facilities were in areas that were nonsensitive. 3M and Weyerhaeuser were subject to less stringent regulation than Merck, whose plant was subject to special restrictions because it was adjacent to a national park, and Intel, whose facility was in Phoenix, a nonattainment area that experiences high levels of automobile pollution. Nevertheless, there should be a cap on how much increased pollution should be allowed by facilities that are in nonsensitive areas (for example, by no more than 30%).

4. Nonsensitive Areas

For pollutants permitted by law and regulation to increase in nonsensitive areas, the main method for computing SEP might be pollutant releases per unit of production. This type of goal will encourage efficiency and provide an incentive for pollution prevention. The amount of the decrease for which a facility should aim could be explicitly spelled out in a *Federal Register* Notice. It should not be left to EPA's discretion. A facility might aim for reductions in pollution per unit of production of at least 20% in one medium or at least 25% if emissions from more than one medium are combined. Per-unit-of-production measures can be difficult to adopt in fast-changing industries that most need regulatory flexibility.

Though there always will be a dispute about what a unit of production is, 3M in its final project agreement was willing to aim for pollution reductions per unit of production of this magnitude. If a facility cannot meet goals for reductions per unit of production, it should be able to petition EPA to approve its permit on the basis of other factors (for example, the potential for a technological breakthrough). The burden of proof would be on the facility to demonstrate technological superiority. The facility must show EPA that what it is doing constitutes a technological breakthrough that, if successful, can be applied more widely among facilities in an industry. The potential technological breakthrough should help solve an environmental problem that EPA considers fairly critical, as was true in the Weyerhaeuser case.

5. Sensitive Areas

In sensitive areas such as those where Intel or Merck were located, the criteria for pollution reductions per unit of production would not hold. In these areas, there should be decreases in actual pollution levels. Again, however, we believe that these decreases should not be decided primarily on a case-by-case basis. The number involved could be spelled out in a *Federal Register* Notice. Using Merck as a precedent, a facility in a sensitive area might have to achieve actual pollution reductions of at least 20%—or perhaps 25% if emissions from more than one medium are combined. If, as with Intel, the unit is still under construction and no baseline exists, the facility applying for the permit should find a comparable facility, preferably owned by its parent company. Or, if such a comparable facility does not exist, the firm should use one owned by another company as the basis for calculations. If there is no comparable facility anywhere, then the approval of an XL permit should be based on our first criterion, allowed releases.

6. Baselines

A key issue is the baseline that would be used in making these determinations: it needs to be consistent across cases. It would be best to spell out the baseline in a *Federal Register* Notice so that it would not be a source of controversy (Stewart 2001). On the basis of the precedents established in our cases, the baseline might be three-year averages for the period before the XL permit goes into effect. The arguments that Merck and Weyerhaeuser made were that this type of baseline accommodates the possibility of unusual fluctuations in either pollution or production in the year immediately preceding an XL permit. Thus, it makes sense to use such a baseline that extends back in time and does not rely exclusively on what might have been atypical annual events.

7. An Extended Period

When does a facility reach its goal of, say, a 25% reduction in air pollutant emissions per unit of production? An extended period for the life of the project and permit allows businesses to justify substantial investments in innovative approaches. A 10-year life for XL permits would allow managers to avoid having to operate under what 3M called a dual regulatory system and also allow them to take advantage of business investment in innovative techniques. What are the intermediate goals until then? Again, site-specific, industry-specific features appear. Some cases are amenable to continuous improvement—for instance, a 50% reduction on average every two years. Other cases may require five years before any improvement is apparent. Changes in the permit should be allowed if new studies or information come to light that show substantial risk to human health or the environment. Over the life of the permit as regulations are modified, the releases a facility is allowed are likely to change. In the 3M case, because the stringency of existing requirements was likely to go up, EPA felt constrained from establishing a 10-year permit.[1]

Thus, XL permits probably should be subject to some midstream reevaluation (for example, after five years). For the permit to be renewed, specific standards should be in place. They could be explicitly spelled out in a *Federal Register* Notice indicating how much more stringent than current allowable releases the XL limits should be. For instance, as long as the XL limits remain 20% more stringent than current allowable emissions in at least one environmental medium (air, water, or land)—or perhaps more than 25% more stringent, if emissions from more than one medium are combined—the permit should be renewed for another five years. Of course, if special circumstances were involved, an applicant could petition EPA to extend its XL permit even if it did not meet this standard. The burden of proof would be on the applicant to demonstrate why it needed the exception. If, after 10 years, the facility continued to meet its XL requirements, the permit, subject to further review, should be considered for renewal.

8. Economic Criteria

In petitioning for an XL permit, a company may make an economic argument. But this type of argument should not be a central determining factor in EPA's deliberations. Economic arguments can supplement environmental arguments and may tilt EPA decisionmakers in close-call cases. If a project sponsor comes forth with an idea that would substantially reduce the cost of regulation, EPA should pursue the idea as a special case even if the direct environmental benefits are less compelling. However, in general, the main criteria EPA should use in judging an XL proposal would be environmental, such as those already proposed:

1. the difference between the limit in the proposed XL permit and current allowed emissions;

2. reductions in actual pollution, or in pollution per unit of production, depending on whether the pollutants are highly toxic and hazardous and the facility is in a sensitive or nonsensitive area; and
3. the perceived value of the regulatory innovations to be tested by the project.

Conclusion

Though this proposal has many positive features, the resulting structure is by no means a panacea. It is likely to require variances, exceptions, and judgements by EPA or other stakeholders. If a lesson has emerged from the cases examined in this book, it is that pollution control is site specific and exceptions to one-size-fits-all rules for SEP—such as discussions about the relative toxicity of pollutants—are sure to arise.

Indeed, people in the business community may object to this proposal. They might perceive the SEP criteria as being complex and risky. They might say that they should not have to demonstrate SEP to achieve regulatory flexibility. Flexibility simply should be granted to them because they are deserving of reasonable regulation. It undoes aspects of existing laws and statutes that are not sound, and regulated parties should not have to bear an extra burden of SEP for sensible rules. People in the environmental community, conversely, may argue that definitions of SEP based on the precedents established in these cases may be too lax; they let businesses too easily off the hook. We would counter that the definitions that emerge from the cases in this book show what responsible companies such as Intel, Merck, 3M, and Weyerhaeuser can do. The potential for achieving environmental benefits systemwide outweigh the risks that might arise from individual pilots, if they are approved based on these requirements.

Note

[1]Stringency is likely to grow for a number of reasons. Some sites have grandfathered-in units—they are not currently subject to regulation because they are in attainment areas for achieving national air quality standards and were in place prior to passage of the Clean Air Act. As these units are replaced, significant changes are made, or the status of the areas where they are found changes, they will be subject to regulation that is substantially more stringent than the regulation to which they are currently subject. 3M had a number of units in this category. Stringency also may go up because new laws are passed or new regulations are brought into existence in response to scientific discoveries or technological breakthroughs.

References

Abbott, A. 1990. A Primer on Sequence Methods. *Organization Science* 1(4): 375–92.
———. 1992. From Causes to Events. *Sociological Methods and Research* 20(4): 428–55.
Abt Associates. 1996. Risk Benefits for Project XL: An Evaluation of Baseline Conditions at 3M's Hutchinson, Minnesota, Facility. Report prepared for Office of Policy Planning and Evaluation, U.S. EPA. Contract 68-C4-0060. Washington, DC: U.S. EPA.
Air and Waste Management Association. 1996a. *New Source Review*. Report prepared for U.S. EPA Office of Air Quality Planning and Standards. M. Johnson, G. McCutchen, and K. Weiss, technical chairs. Pittsburgh: Air and Waste Management Association.
———. 1996b. Title 5: Compliance vs. Production. *EM* Sept.: 1–64.
Arrow, K., R. Mnookin, L. Ross, A. Teversky, and R. Wilson. 1995. *Barriers to Conflict Resolution*. New York: W.W. Norton.
Arthur, B. 1997. *Increasing Returns and Path Dependence in the Economy*. Ann Arbor, MI: University of Michigan Press.
Arthur D. Little. 1996. Sustainable Industrial Development: Sharing Responsibilities in a Competitive World. Conference paper prepared for Ministry of Housing, Spatial Planning, and the Environment and for Ministry of Economic Affairs, February, The Hague, Netherlands.
Aspen Institute. 1996. The Alternative Path: A Cleaner, Cheaper Way to Protect and Enhance the Environment. Paper prepared by Program on Energy, the Environment, and the Economy. Queenstown, MD: Aspen Institute.
Axelrod, R. 1984. *The Evolution of Cooperation*. New York: Basic Books.

Bacow, L., and M. Wheeler. 1984. *Environmental Dispute Resolution*. New York: Plenum.

Bazerman, M., J. Baron, and K. Shonk. 2001. *"You Can't Enlarge the Pie": The Psychology of Ineffective Government*. New York: Basic Books.

Bazerman, M., J. Curhan, and D. Moore. 2000. The Death and Rebirth of the Social Psychology of Negotiation. In *Blackwell Handbook of Social Psychology*, edited by M. Hewstone and M. Brewer. London: Blackwell.

Bingham, G. 1986. *Resolving Environmental Disputes: A Decade of Experience*. Washington, DC: Conservation Foundation.

Blackman, A., and J. Mazurek. 1999. The Cost of Developing Site-Specific Environmental Regulations: Evidence from EPA's Project XL. Discussion Paper 99-35. Washington, DC: Resources for the Future.

Bower, J., and T. Hout. 1988. Fast-Cycle Capability for Competitive Power. *Harvard Business Review* Nov.–Dec.: 110–18.

Breyer, S. 1993. *Breaking the Vicious Circle: Toward Effective Risk Regulation*. Cambridge, MA: Harvard University Press.

Brodt, S., and L. Dietz. 1999. Shared Information and Information Sharing: Understanding Negotiation as Collective Construal. *Research on Negotiation in Organizations* 7: 263–83.

Browner, C. 1998. Remarks made at Information-Based Environmental Regulation: The Beginning of a New Regulatory Regime? A conference organized at Columbia University Law School, Oct. 29–30, New York.

Bryner, G. 1996. Reforming the Regulatory Process: Congress and the Next Generation of Environmental Laws. Paper presented at Annual Meeting of Western Political Science Association, March 14–16, San Francisco.

Buelow, C. 1996. An Experiment in Flexible Permitting: Project XL. Master's project, Nicholas School of the Environment, Duke University, Durham, NC.

———. 1997. Barriers to Regulatory Reform as Experienced in the 3M Project XL Pilot. Project submitted for master's in environmental management degree, Duke University, Durham, NC.

Canon, J. 1999. Bargaining, Politics, and Law in Environmental Regulation. Paper prepared for discussion at Wharton School Conference on Environmental Contracts and Regulatory Innovation, Sept. 24–25, Philadelphia.

CBEE (Collaborating for a Better Environment and Economy in Minnesota). 1995. *Draft Summary of P2 Dialogues 1–7, 1994–1995*. Minneapolis: Minnesota Environmental Initiative.

———. 1996. *Recommendations of the Pollution Prevention Dialogue*. Minneapolis: Minnesota Environmental Initiative.

Chertow, M., and D. Esty (eds.). 1997. *Thinking Ecologically*. New Haven, CT.: Yale University Press.

Clayton, A., and N. Radcliffe. 1996. *Sustainability*. Boulder, CO: Westview Press.

Clinton, B. 1995. Remarks by the President on Project XL, Nov. 3, White House, Washington, DC. http://www.govinfo.library.unt/npr/library/speeches/26621.html (accessed May 19, 2002).

Clinton, B., and A. Gore. 1995. *Reinventing Environmental Regulation*. March 16, Washington, DC. http://www.govinfo.library.unt.edu/npr/library/rsreport/251a.html (accessed May 19, 2002).

Coglianese, C. 1997. Assessing Consensus: The Promise and Performance of Negotiated Rulemaking. *Duke Law Journal* 46: 1255–1349.

———. 1999. The Limits of Consensus in Environmental Regulation. Paper prepared for discussion at Wharton School Conference on Environmental Contracts and Regulatory Innovation, Sept. 24–25, Philadelphia.

Coglianese, C., and J. Nash. 2001. *Regulating from the Inside: Can Environmental Management Systems Achieve Policy Goals?* Washington, DC: Resources for the Future.

Conlon, D., and D. Sullivan. 1999. Intractable Disputes Involving Organizations. *Research on Negotiation in Organizations* 7: 141–75.

Cushman, D., and S.S. King. 1991. High Speed Management: A Revolution in Organizational Communication in the 1990s. *Communication Yearbook* 16: 209–36.

Davies, J., and J. Mazurek. 1996. *Industry Incentives for Environmental Improvement: Evaluation of U.S. Federal Initiatives.* Report prepared for Global Environmental Management Initiative. Washington, DC: Resources for the Future.

Davis, K. 1977. *Discretionary Justice.* Urbana, IL: University of Illinois Press.

Dawson, E. 1998. Looking at Voluntary Participation Programs: A Case Study of Project XL at the Weyerhaeuser Flint River Facility. Senior honor's thesis, Natural Resources and the Environment, University of Michigan, Ann Arbor.

Delmas, M., and J. Mazurek. 2001. Alliances with Governments: A Transaction Costs Perspective (The Case of the XL Program). Working paper, University of California, Santa Barbara.

Delmas, M., and A. Terlaak. 2000. A Framework for Analyzing Environmental Voluntary Agreements. *California Management Review* 43(3): 44–63.

Dews, F. 1997. A Project XL Pilot Summary: 3M, Hutchinson, Minnesota, July 14 Draft. Washington, DC: Government Studies Program, Brookings Institution.

Dorf, M., and C. Sabel. 1998. A Constitution of Democratic Experimentalism. *Columbia Law Review* 98(2): 267–473.

Dougherty, J., and R. Pfaltzgraff. 1970. *Contending Theories of International Relations.* New York: J. B. Lippincott.

Dumaine, B. 1989. How Managers Can Succeed Through Speed. *Fortune* Feb. 13: 54–59.

ECOS (Environmental Council of the States). 1997a. Draft *Federal Register* Notice, Joint EPA/State Agreement to Pursue Regulatory Innovation, Oct. 29.

———. 1997b. Draft Joint EPA/State Agreement to Pursue Regulatory Innovation, Feb. 10. Reprinted in *State Environmental Monitor*, Feb. 14.

———. 1997c. ECOS Proceedings of the Environmental Regulatory Innovations Symposium, Environmental Regulatory Innovations Symposium, Nov. 5–7, Minneapolis.

———. 1997d. Report of Innovation Breakout Groups, Environmental Regulatory Innovations Symposium, Nov. 5–7, Minneapolis.

Eisenhardt, K. 1989. Building Theories from Case Study Research. *Academy of Management Review* 14(4): 532–50.

Eisenhardt, K., and B. Tabrizi. 1995. Accelerating Adaptive Processes: Product Innovation in the Global Computer Industry. *Administrative Science Quarterly* 40: 84–110.

Environmental Resources Management Group. 1993a. Clean Air Act Primer, May. Walnut Creek, Texas.

————. 1993b. The 1990 Clean Air Act Amendments, Charting a Compliance Course. Prepared by Weiss, Gallagher, and Klaber, April, Walnut Creek, Texas.

Faegre & Benson. 1994. Nowhere to Run, Nowhere to Hide: Operating Permits under the Clean Air Act of 1990. Minneapolis, MN.

Fidler, L., and J. Johnson. 1984. Communication and Innovation Implementation. *Academy of Management Review* 9(4): 704–11.

Fiorino, D. 1995. *Making Environmental Policy.* Los Angeles: University of California Press.

Fischbeck, P., and R.S. Farrow (eds.). 2001. *Improving Regulation: Cases in Environment, Health, and Safety.* Washington, DC: Resources for the Future.

Fisher, R. 1994. *Beyond Machiavelli: Tools for Coping with Conflict.* New York: Penguin.

Fisher, R., W. Ury, and B. Patton. 1991. *Getting to Yes.* New York: Penguin.

Foray, D., and A. Grubler. 1996. Introduction to Special Issue on Technology and the Environment. *Technological Forecasting and Social Change* 53(1): 3–13.

Freeman, C. 1996. The Greening of Technology and Models of Innovation. *Technological Forecasting and Social Change* 53(1): 27–39.

Freeman, J. 1997. Collaborative Governance in the Administrative State. *UCLA Law Review* 45(1): 1–98.

Funk, W. 1986. Bargaining toward the New Millennium: Regulatory Negotiation and the Subversion of the Public Interest. *Duke Law Journal* 46: 1351–88.

Gersick, C. 1991. Revolutionary Change Theories: A Multilevel Exploration of the Punctuated Equilibrium Paradigm. *Academy of Management Review* 16: 10–36.

Ginsberg, G., and C. Cumis. 1996. EPA's Project XL: A Paradigm for Promising Regulatory Reform. *ELR News and Analysis* 26: 10059–64.

Gould, S. 1991. *Bully for Brontosaurus.* New York: W.W. Norton.

————. 1996. The Panda's Thumb of Technology. In *Managing Strategic Innovation and Change*, edited by M. Tushman and P. Anderson. New York: Oxford University Press, 68–75.

Graham, M. 1998. Environmental Protection and the States. *Brookings Review* 16(1): 22–26.

Gray, B. 2000. Freeze Framing: The Timeless Dialogue of Intractability Surrounding Voyageurs National Park. Paper prepared for Academy of Management. Toronto: Academy of Management.

Grazman, D., and A. Van de Ven. 1992. Building an Event Sequence File. Working paper. University of Minnesota, Carlson School of Management, Minneapolis, MN.

Gregory, R., T. McDaniels, and D. Fields. 2001. Decision Aiding, Not Dispute Resolution: Creating Insights through Structured Environmental Decisions. *Journal of Policy Analysis and Management* 20(3): 415–33.

Grin, J., and H. Van Den Graaf. 1996. Implementation as Communicative Action. *Policy Sciences* 29: 291–319.

Gundling, E. 2000. *The 3M Way to Innovation: Balancing People and Profit.* Tokyo: Kodansha.

Guth, W., and I. MacMillan. 1986. Strategy Implementation versus Middle Management Self-Interest. *Strategic Management Journal* 7: 313–27.

Hirsch, D. 1998. Bill and Al's XL-ENT Adventure: An Analysis of the EPA's Legal Authority to Implement the Clinton Administration's Project XL. *University of Illinois Law Review* 1998(1): 129–72.

————. 2001. Project XL and the Special Case: The EPA's Untold Success Story. *Columbia Journal of Environmental Law* 26(2): 101–44.

Hoffman, A., H. Riley, J. Troast, and M. Bazerman. 2002. Cognitive and Institutional Barriers to New Forms of Cooperation on Environmental Protection: Insights from Project XL and Habitat Conservation Plans. *American Behavioral Scientist* 45(5): 820–845.

Hosmer, L. 1995. Trust. *Academy of Management Review* 20(2): 379–403.

Howard, D. 1994. *The Death of Common Sense*. New York: Random House.

John, D. 1994. *Civic Environmentalism: Alternatives to Regulation in States and Communities*. Washington, DC: CQ Press.

Kessler, E., and A. Chakrabarthi. 1996. Innovation Speed: A Conceptual Model of Context, Antecedents, and Outcomes. *Academy of Management Review* 21(4): 1143–91.

Kramer, R.M. 1991. The More the Merrier? Social Psychological Aspects of Multiparty Negotiations in Organizations. In *Research on Negotiation in Organizations*, edited by R.J. Bies, R.J. Lewicki, and B.H Sheppard. Greenwich, CT: JAI, vol. 3, 307–32.

Kwon, S., and L. Weingart. 2000. When You Get a Unilateral Concession from the Other Party. Paper prepared for Academy of Management Conference, August 4-9, Toronto.

Lax, D., and J. Sebinius. 1986. *The Manager as Negotiator*. New York. Free Press.

Landy M., M. Roberts, and S. Thomas. 1990. *The Environmental Protection Agency: Asking the Wrong Questions*. New York: Oxford University Press.

Lewicki, R., and J. Litterer. 1985. *Negotiation*. Homewood, IL: Irwin.

Long, F., and M. Arnold. 1995. *The Power of Environmental Partnerships*. Fort Worth, TX: Dryden Press.

Marcus, A. 1980. *Promise and Performance: Choosing and Implementing an Environmental Policy*. Westport, CT: Greenwood Press.

————. 1991. EPA's Organizational Structure. *Law and Contemporary Problems* 54(4): 5–41.

Marcus, A., D. Geffen, S. Erickson, A. Miller, and B. Smith. 1998. Collaborating for a Better Environment and Economy in Minnesota: Progress Report, February 1, 1997–December 31, 1997. University of Minnesota, Minneapolis, MN.

Marcus, A., D. Geffen, A. Frisch, A. Miller, and S. Smith 1997. Collaborating for a Better Environment and Economy in Minnesota: Progress Report, August 1, 1996–January 31, 1997. University of Minnesota, Minneapolis, MN.

Marcus, A., D. Geffen, K. Sexton, and B. Smith. 1995. Advising, Monitoring, and Evaluating a Minnesota Pollution Control Agency Pilot Project for Flexible, Multi-Media Permitting. Proposal to the EPA. University of Minnesota, Minneapolis, MN.

————. 1996a. Advising, Monitoring, and Evaluating a Minnesota Pollution Control Agency Pilot Project for Flexible, Multi-Media Permitting: Progress Report, June 15, 1996–October 31, 1996. University of Minnesota, Minneapolis, MN.

————. 1996b. Advising, Monitoring, and Evaluating a Minnesota Pollution Control Agency Pilot Project for Flexible, Multi-Media Permitting: Progress Report, March 1, 1996–June 15, 1996. University of Minnesota, Minneapolis, MN.

————. 1997. Advising, Monitoring, and Evaluating a Minnesota Pollution Control Agency Pilot Project for Flexible, Multi-Media Permitting: Progress Report, November 1, 1996–April 30, 1997. University of Minnesota, Minneapolis, MN.

Marcus, A., D. Geffen, K. Sexton, and C. Wiessner. 1998. Advising, Monitoring, and Evaluating a Minnesota Pollution Control Agency Pilot Project for Flexible, Multi-Media Permitting: A Review of Our Research Group's Efforts Over the Last 12 Months. University of Minnesota, Minneapolis, MN.

Marcus, A., M. Nadel, and K. Merrikin. 1984. The Applicability of Regulatory Negotiation to Disputes Involving the Nuclear Regulatory Commission. *Administrative Law Review* 36(3): 213–38.

Mayer, R., and J. Davis. 1995. An Integrative Model of Organizational Trust. *Academy of Management Review* 20(3): 709–34.

Michael, D. 1996. Cooperative Implementation of Federal Regulations. *Yale Journal of Regulation* 13(2): 537–600.

Minnesota State Legislature. 1996. *The Environmental Regulatory Innovations Act.* April 3. Minn. Stat. S 114C.01-15. Saint Paul: Minnesota State Legislature.

Mohin, T. 1997. The Alternative Compliance Model: A Bridge to the Future of Environmental Management. *ELR News and Analysis* 27: 10345–56.

MPCA (Minnesota Pollution Control Agency). 1995a. Facts About Project XL, Air Quality Division. November. Saint Paul: MPCA.

———. 1995b. Minnesota Project XL Proposal Submitted to the U.S. EPA. June 16. Saint Paul: MPCA.

———. 1995c. Project XL MOU: Draft Memorandum of Understanding Delegating the Project XL Program between the U.S. EPA and the MPCA. Dec. 21. Saint Paul: MPCA.

———. 1996a. The Draft Permit. May 2. Saint Paul: MPCA.

———. 1996b. Findings of Fact and Conclusions in the Matter of the Decision on the Need for an Environmental Impact Statement for the Proposed 3M Hutchinson Project XL. June 4. Saint Paul: MPCA.

———. 1996c. Minnesota XL–Permit No. 96-01 to 3M Company. Final draft, May 14. Saint Paul: MPCA.

———. 1996d. Project XL MOU: Draft Memorandum of Understanding Delegating the Project XL Program Between the U.S. EPA and the MPCA. Jan. 16. Saint Paul: MPCA.

———. 1996e. Project XL MPCA Management Report. March 8. Saint Paul: MPCA.

———. 1996f. Public Notice of Proposed Issuance of Minnesota XL Permit and Revocation of the MPCA Permit to be Replaced by the Minnesota XL Permit, and Minnesota XL–Final Project Agreement (FPA) between the MPCA, EPA, and 3M. Draft. May 29. Saint Paul: MPCA.

———. 1998. 1998 Project XL Report to the Minnesota Legislature. Saint Paul: MPCA.

Muthoo, A. 1999. *Bargaining Theory with Application.* Cambridge, U.K.: Cambridge University Press.

Nash, J. 2001. Tiered Environmental Regulation: Lessons from the StarTrack Program. Paper presented at Workshop on Voluntary, Collaborative, and Information Based Policies: Lessons and Next Steps for Environmental and Energy Policy in the United States and Europe, Harvard University, May 10–11, Cambridge, MA.

NAPA (National Academy of Public Administration). 1995. *Setting Priorities Getting Results: A New Direction for the EPA.* Washington, DC: NAPA.

———. 1997. *Resolving the Paradox of Environmental Protection.* Washington, DC: NAPA.

Neale, M., and M. Bazerman. 1991. *Cognition and Rationality in Negotiation*. New York: Free Press.

Odell, J. 1999. The Negotiation Process and International Economic Organizations. Presentation to American Political Science Association, September 3, Atlanta.

O'Leary, R. 1995. Environmental Mediation: What Do We Know and How Do We Know It? In *Mediating Environmental Conflicts: Theory and Practice*, edited by J.W. Blackburn and W. Bruce. Westport, CT: Quorum Books, 17–37.

O'Leary, R., R. Durant, D. Fiorino, and P. Weiland. 1999. *Managing for the Environment*. San Francisco: Jossey Bass.

Orts, E. 1995a. Reflexive Environmental Law. *Northwestern University Law Review* 89(4): 1227–1340.

Orts, E. 1995b. Simple Rules and the Perils of Reductionist Legal Thought. *Boston University Law Review* 75(5): 1441–79.

Orts, E., and P. Murray. 1997. Environmental Disclosure and Evidentiary Privilege. *University of Illinois Law Review* 1: 1–69.

Osborne, D., and T. Gabler. 1993. *Reinventing Government*. New York: NAL/Dutton.

Palmer, L.G., and L. Thompson. 1995. Negotiation in Triads: Communication Constraints and Tradeoff Structure. *Journal of Experimental Psychology* 1: 83–94.

Pedersen, W. 1995. Can Site-Specific Pollution Controls Furnish an Alternative to the Current Regulatory System and a Bridge to a New One? *Environmental Law Reporter* 25: 10486–90.

Percival, R., and D. Alevizatos. 1997. *Law and the Environment: A Multidisciplinary Reader*. Philadelphia: Temple University Press.

Pettigrew, A. 1985. Examining Change in the Long-Term Context of Culture and Politics. In *Organizational Strategy and Change*, edited by J. Pennings and associates. San Francisco: Jossey Bass, 269–317.

PPC (Pilot Project Committee). 1995. Project XL Measurements Table. Draft, Oct. 31. Minneapolis, MN.

President's Council on Sustainable Development. 1996. Sustainable America. http://clinton2.nara.gov/PCSD/Publications/TF_Reports/amer-top.html (accessed July 2, 2002).

Public Interest Representatives. 1995. Joint Statement on the Clinton Administration's Project XL Announcement. Nov. 3. Washington, DC: Public Interest Representatives.

———. 1996. Alternative Regulatory Pathway: Evaluation Criteria. October. Washington, DC: Public Interest Representatives.

Raiffa, H. 1982. *The Art and Science of Negotiation*. Cambridge, MA: Belknap Press.

Ring, P., and A. Van de Ven. 1992. Structuring Cooperative Relationships Between Organizations, *Strategic Management Journal* 13: 483–98.

———. 1994. Developmental Processes of Cooperative Interorganizational Relationships. *Academy of Management Review* 19(1): 90–118.

Ronchak, A. 1995. Project XL Overview. Paper presented to M2P2 Permitting Project Conference, Nov. 8, Baltimore.

———. 1996. 3M XL Pilot History. December 24. Saint Paul: MPCA.

Rosenbaum, W. 1998. *Environmental Politics and Policy*, 4th ed. Washington, DC: CQ Press.

Ross L., and C. Stillinger. 1991. Barriers to Conflict Resolution. *Negoiations Journal* 7: 389–404.

Ruckelshaus, W. 1996. Trust in Government: A Prescription for Restoration. Webb Lecture, National Academy of Public Administration, Nov. 15, Washington, DC.

Ruttan, S. 1996. Induced Innovation and Path Dependence. *Technological Forecasting and Social Change* 53(1): 41–59.

Sabel, C. 1991. Studied Trust: Building New Forms of Cooperation in a Volatile Economy. In *Industrial Districts and Local Economic Regeneration*, edited by Frank Pyke and Werner Sengenberger. Geneva: International Institute for Labour Studies, 215–250.

Schilling, M., and C. Hill. 1998. Managing the New Product Development Process. *Academy of Management Executive* 12(3): 67–82.

Sexton, K., A. Marcus, K.W. Easter, and T. Burkhardt. 1999. *Better Environmental Decisions: Strategies for Governments, Businesses, and Communities.* Washington, DC: Island Press.

Sexton, K., and B.S. Murdock. 1996. *Environmental Policy in Transition: Making the Right Choices.* Vol. 1 in Minnesota Series in Environmental Decision Making. Minneapolis: Center for Environment and Health Policy, School of Public Health, University of Minnesota.

Sexton, K., B.S. Murdock, and A. Marcus. 2002. Cooperative Environmental Solutions: Acquiring Competence for Multi-Stakeholder Partnerships. In *Environmental Agreements: Process, Practice, and Future Use*, edited by P. ten Brink. Sheffield, U.K.: Greenleaf Publishing, 64–81.

Sinsheimer, P., and R. Gottlieb. 1995. Pollution Prevention Voluntarism: The Example of 3M. In *Reducing Toxics: A New Approach to Policy and Industrial Decision Making*, edited by R. Gottlieb. Washington, DC: Island Press, 389–429.

Skrzycki, C. 1996. Critics See a Playground for Polluters in EPA's XL Plan. *Washington Post*, Jan. 24, D1.

Smith, K., S. Carroll, and S. Ashford. 1995. Intra- and Interorganizational Cooperation: Toward a Research Agenda. *Academy of Management Journal* 38: 7–24.

Sparrow, C. 1998. Regulatory Reform: Putting the Pieces Together. *State Environmental Monitor* Feb. 2: 30–34.

Spence, D., and L. Gopalakrishnan. 1999. Bargaining Theory and Regulatory Reform: The Political Logic of Inefficient Regulation. Working paper. Nashville: Vanderbilt University Law School.

Steinzor, R. 1996. Regulatory Reinvention and Project XL: Does the Emperor Have Any Clothes? *ELR News and Analysis* 26: 10527–37.

Stewart, R. 2001. A New Generation of Environmental Regulation? *Capital University Law Review* 29(1): 21–183.

Stewart, T. 1996. 3M Fights Back. *Fortune* 133(2): 94–102.

Susskind, L., and J. Cruikshank. 1987. *Breaking the Impasse.* New York: Basic Books.

Susskind, L., P. Levy, and J. Thomas-Larmer. 2000. *Negotiating Environmental Agreements.* Washington, DC: Island Press.

Susskind, L., and J. Secunda. 1999. The Risks and the Advantages of Agency Discretion: Evidence from EPA's Project XL. *UCLA Journal of Environmental Law and Policy* 17(1): 67–116.

Thorvig, L. 1996. Minnesota's Regulatory Flexibility Project. In *Environmental Policy in Transition*, edited by K. Sexton and B. Murdock. Minneapolis: Center for Environment and Health Policy, School of Public Health, University of Minnesota, 39–46.

3M Company. No date. 3M's Pollution Prevention Pays. Saint Paul: 3M Company.

———. 1990a. Corporate Environmental Policy. Saint Paul: 3M Company.

———. 1990b. Pollution Prevention Pays: Status Report. Saint Paul: 3M Company.

———. 1991a. Air Emissions Reduction Program. Saint Paul: 3M Company.

———. 1991b. Early Reductions of Hazardous Air Pollutants. Saint Paul: 3M Company.

———. 1991c. 3M and the Environment: A Progress Report, Volume 1. Saint Paul: 3M Company.

———. 1992. The Environment: A Special Report. Saint Paul: 3M Company.

———. 1995a. The Beyond Compliance Emissions Reduction Act of 1994. Saint Paul: 3M Company.

———. 1995b. The Beyond Compliance Emissions Reduction Act of 1995: Executive Summary. Saint Paul: 3M Company.

———. 1996a. The Covenant Draft. Jan. 16. Saint Paul: 3M Company.

———. 1996b. What Is Superior Environmental Performance? July 25. Saint Paul: 3M Company.

Townsend, A., S. DeMarie, and A. Hendrickson. 1998. Virtual Teams. *Academy of Management Executive* 12(3): 17–30.

Tushman, M., and P. Anderson (eds.). 1997. *Managing Strategic Innovation and Change.* New York: Oxford University Press.

U.S. EPA (Environmental Protection Agency). 1994. Air Toxin Rule for the Magnetic Tape Manufacturing Industry. http://www.epa.gov/ProjectXL/3mhut/0098.htm (accessed May 20, 2002).

———. 1995a. Framework for Memorandum of Understanding with Minnesota Pollution Control Agency, Oct. 30.

———. 1995b. Regulatory Reinvention (XL) Pilot Projects, Solicitation of Proposals and Request for Comments, Environmental Protection Agency. *Federal Register* Notice FRL 5197-9: 27283–90, May 23.

———. 1996a. 3M: Hutchinson, MN. http://www.epa.gov/ProjectXL/3mhut/index.htm (accessed May 20, 2002).

———. 1996b. Innovative Environmental Technologies Notice: Solicitation of Proposals for and Request for Comment on Project XL. http://www.epa.gov/ProjectXL/eval4.htm (accessed May 20, 2002).

———. 1996c. Minnesota Project XL Proposal: Background. http://www.epa.gov/ProjectXL/eval4.htm (accessed May 20, 2002).

———. 1996d. *National Stakeholder Involvement in Project XL: Some Questions and Answers, and Project XL Communities, XL Community Pilot Program.* Washington, DC: U.S. EPA.

———. 1996e. Partnerships in Preventing Pollution: A Catalogue of the Agency's Partnership Programs. Office of the Administrator. EPA 100-B-96-001. Washington, DC: U.S. EPA.

———. 1996f. The Plain English Guide to the Clean Air Act. Office of Air Quality Planning and Standards. http://www.epa.gov/oar/oaqps/peg_caa/pegcaain.html (accessed May 20, 2002).

———. 1996g. Project XL Project Development Process Guidelines for Stakeholder Involvement. Draft. Washington, DC: U.S. EPA.

———. 1996h. *Explore an XL Project*. http://www.epa.gov/projectxl/explorxl.htm (accessed May 20, 2002).

———. 1997a. Environmental Protection Agency, Regulatory Reinvention (XL) Pilot Projects, Notice of Modifications to Project XL. *Federal Register* Notice FRL-5811-7: 19872–82, April 23.

———. 1997b. *Managing for Better Environmental Results: A Two-Year Anniversary Report on Reinventing Environmental Protection*. Office of the Administrator. EPA 100-R-97-004. Washington, DC: U.S. EPA.

———. 1997c. New Directions: A Report on Regulatory Reinvention. Office of the Administrator. EPA100-R-97-001. Washington, DC: U.S. EPA.

———. 1997d. *Regulatory Plan and Semiannual Agenda of Regulatory and Deregulatory Actions*. Office of Policy, Planning, and Evaluation. EPA 230-Z-97-002. Washington, DC: U.S. EPA.

———. 1997e. XL Implementation and Evaluation. http://www.epa.gov/projectxl/implemen.htm (accessed May 20, 2002).

———. 1997f. XL Site Specific Rulemaking for Merck & Co., Stonewall Plant, Part 3. Environmental Protection Agency. *Federal Register* 40, parts 52, 60, 264, and 265.

———. 1998a. *The Changing Nature of Environmental and Public Health Protection*. Office of Reinvention. EPA 100-R-98-003. Washington, DC: U.S. EPA.

———. 1998b. *Guide for EPA XL Project Teams, Draft*. Sept. 18. Washington, DC: U.S. EPA.

———. 1998c. *New Directions: A Report on Regulatory Reinvention*. EPA 100-R-98-04. Washington, DC: U.S. EPA.

———. 1998d. Project XL: Best Practices for Proposal Development: A Guide to Help Project Sponsors Create Effective XL Proposals. Working paper. Office of Reinvention. Washington, DC: U.S. EPA.

———. 1998e. *Project XL Preliminary Status Report*, Office of the Administrator. EPA-100-R-98-008. Washington, DC: U.S. EPA.

———. 1998f. *Reinventing Environmental Protection—EPA's Approach*. Office of Reinvention. EPA 1000F-98-010. Washington, DC: U.S. EPA.

———. 1998g. *Review Draft Project XL Stakeholder Involvement Guide*. Office of Reinvention. Washington, DC: U.S. EPA.

———. 1998h. Solicitation of Additional Projects under Project XL, EPA. *Federal Register* Notice, June 12. Also, http://www.epa.gov/fedrgstr/EPA-WASTE/1998/June/Day-23/f16398.htm (accessed May 20, 2002).

———. 1998i. Project XL Reengineering Efforts and Results. http://www.epa.gov/projectxl/xlimpr_b.htm (accessed May 20, 2002).

———. 1998j. Project XL Process Improvements. http://www.epa.gov/projectxl/xlimprov.htm (accessed May 20, 2002).

———. 1999a. Project XL: Best Practices for Proposal Development. http://www.epa.gov/projectxl/eval9.htm (accessed May 20, 2002).

———. 1999b. XL Project Progress Report: Weyerhaeuser Flint River Operations. Office of Reinvention. EPA/100F-99-004. Washington, DC: U.S. EPA.

Van de Ven, A. 1992. Suggestions for Studying Strategy Process. *Strategic Management Journal* 13: 169–88.

Volokh, A., L. Scarlett, and S. Bush. 1998. Race to the Top: The Innovative Face of State Environmental Management. Reason Public Policy Institute, National Environmental Policy Institute, Washington, D.C. Also, http://www.rppi.org/ps/ps239.htm (accessed May 20, 2002).

Weber, E. 1998. *Pluralism by the Rules.* Washington, DC: Georgetown University Press.

Weick, K. 1979. *Social Psychology of Organizing.* Reading, MA: Addison Wesley.

Wilson, J. 1989. *Bureaucracy.* New York: Basic Books.

Wood, J., and T. Colosi. 1996 "The Subtle Art of Negotiation," "It Takes More than Two to Argue," and "Managing Negotiations," *London Financial Times Mastering Management Series*, January (9, 10, 11): 419–442.

Yin, R. 1989. *Case Study Research: Design and Methods.* Newbury Park, CA: Sage Publications.

Acknowledgements

This book would not have been possible without a grant (R824754) from EPA. To write the book, we drew on the report we wrote for the agency, *Impasse in the Movement toward a New Competence in Environmental Management* (1999). Katherine Dawes, Chris Knopes, and others at EPA commented on the report. We were direct participants in some of the events described and also interviewed many of the participants. In writing this book, we have relied heavily on the notes from the meetings we attended, the transcripts of interviews that we taped, and the many documents we collected.

We are especially grateful to people we interviewed, some of whom we talked with again and again. Among them, we especially thank Lisa Thorvig, Andy Ronchak, Peggy Bartz, and Al Inness of the Minnesota Pollution Control Agency; Tom Zosel, Dave Sonstegard, Dave Wefring, Patty Kirchoff, and Cheryl Kedrowski of the 3M Company; Marilou Martin, Bill Wagner, Brian Barwick, Genevieve Nearmeyer, Rudy Tenasciaovich, and Rachel Rinehard of EPA Region 5 in Chicago; David Gardiner, Lisa Lund, George Wyeth, Bruce Buckheit, John Fogarty, Linda Lay, Michael Goo, Cynthia Cumis, James Lounsbury, and Dan Fiorino of EPA headquarters in Washington; Fred Hansen, former deputy EPA administrator; and Jon Kessler, former EPA head of regulatory reform.

The interview process started in 1995 and resumed between March 1996 and February 1997, when many of the events described in this book took place. The interviews with people at the Minnesota Pollution Control

Agency were carried out mostly in March and April 1996; at 3M in June and July 1996; at EPA Region 5 in July 1996; and at EPA headquarters in February 1997. Alan Miller assisted us with the interviews conducted in Washington, D.C. Most of the interviews were confidential and of the not-for-attribution variety. In exchange for frank discussion, we promised anonymity—that is, we would not directly associate a name with what a person said.

In addition to interviews listed above, David Hawkins and Chris Von Loben Sells of the Natural Resources Defense Council shared their views of the events that took place. In November and December 2000, we interviewed Tedd H. Jett of Merck & Company, Gary Risner of Weyerhaeuser Company, James N. Larsen of Intel Corporation, and Jo Crumbaker of Maricopa County Environmental Services, Arizona (Phoenix).

Parts of this book were presented in draft form at conferences and reviewed by academic experts, EPA staff, environmental professionals, staff members of nongovernmental organizations, and people in the business community. We especially thank Charles Sabel and Bradley Karkkainen of the Columbia University Law School for inviting us to present a part of this study in draft form at the "Information-Based Regulation: The Beginning of a New Regulatory Regime?" conference at Columbia University on October 29 and 30, 1998. At this conference, we obtained feedback and heard presentations from, among others, Carol Browner, former EPA administrator; Archon Fund of the John F. Kennedy School of Government at Harvard University; and Sanford Lewis of the Good Neighbor Project.

Eric Orts of the Wharton School of Business also gave us the opportunity to present parts of this book in draft form at a research conference held September 24 and 25, 1999, at the University of Pennsylvania. The University of Pennsylvania business and law schools sponsored the conference, whose topic was "Environmental Contracts." (See *Environmental Contracts: Comparative Approaches to Regulatory Innovation in the United States and Europe*, edited by Eric and Kurt Deketellaere and published by Kluwer Law International in 2001.) We heard presentations and obtained feedback from Dennis Hirsch of Capital Law School, Joshua Secunda of EPA Region 1 in Boston, Dan Farber of the University of Minnesota Law School, Michael Baram of Boston University Law School, Jonathan Cannon of the University of Virginia Law School, John Maxwell of the University of Indiana Business School, David Spence of Vanderbilt University Law School, and Rena Steinzor of the University of Maryland Law School. Dennis Hirsch, Eric Orts, and Dan Farber later read drafts of the book and provided many useful suggestions.

A number of other people read parts or the entire book and offered suggestions, including: Cary Coglianese of the Kennedy School; Chris Foreman of the University of Maryland and Brookings Institution; Todd Baldwin of Island Press; Terry Davies of Resources for the Future; Barry Rabe of the

University of Michigan; and Michael Kraft of the University of Wisconsin, Green Bay.

We presented a paper based on material in this book at the Association for Public Policy Analysis and Management's Annual Research Conference, which was held in Seattle on November 2-4, 2000. Vicki Norberg-Bohm of the Kennedy School organized the session. From Vicki and the other presenters, we obtained many useful comments. Vicki then invited us to present a paper at a conference that she and Theo de Bruijn organized at the Kennedy School on May 10 and 11, 2001, "Voluntary, Collaborative, and Information-Based Policies: Lessons and Next Steps for Environmental and Energy Policy in the United States and Europe." At that conference, we received useful comments from Shelley Metzenbaum of the Kennedy School and Leslie Carothers of United Technologies. Shelley then read additional drafts and continued to help us in formulating our arguments. Andy Hoffman of the Boston University School of Management, a participant at this meeting, also read and commented on draft chapters.

Max Bazerman of the Harvard Business School, Pri Shah from the Carlson School of Management at the University of Minnesota, and Cathie Ramus of the University of California Bren School assisted us with advice about negotiations and the negotiations literature. We also thank Richard A. Minard Jr., of the National Academy of Public Administration for the help he provided. Carol Wiessner and Brett Smith were part of the research team that worked on the EPA study. They sat through numerous meetings where the events that took place were discussed, and they offered many valuable insights. Their perspectives, as members of local environmental groups and as participants in the Minnesota Pollution Control Agency's stakeholder group, were particularly enlightening.

The Joyce Foundation and the Environmental Quality Board of the State of Minnesota sponsored some of the activities that made this book possible. The Minnesota Environmental Initiative collaborated with us on a number of project-related activities. The Strategic Management Research Center at the Carlson School and the Center for Environment and Health Policy at the School of Public Health of the University of Minnesota also contributed to the project. We especially thank Sharon Hansen of the Strategic Management Research Center.

Mary Keirstead and Jacob Fine helped edit the book. Katherine McDonald served as an assistant to the researchers and gathered and organized some of the documentary evidence. Both Katherine and Theron Shaw of the Humphrey Institute of Public Affairs at the University of Minnesota wrote research papers on Project XL–Minnesota, which we supervised. We benefited from the work they did and also from Christine Buelow's master's thesis on Project XL–Minnesota, which she wrote at Duke University, partially under our supervision. We obtained from Emily Dawson, at the time an undergraduate at the University of Michigan, a well-researched paper she did on the Weyerhaeuser project, which we also found very useful.

We cannot forget Tom Zosel of 3M's Environmental Engineering and Pollution Control staff. Tom was a dedicated 3M employee who tried very hard to formulate innovative environmental permits at a number of 3M facilities and to reach agreements with EPA. He was passionately dedicated to changing the regulatory system and recasting it in a form that he considered more rational. Sadly, Tom passed away unexpectedly before this project was complete.

We have tried to take into account the helpful written and oral comments made by the many parties who read versions of this study in draft form or heard papers in progress at conferences. We take full responsibility for the final product and any errors of fact or misinterpretations that remain.

Index

About the Authors

Alfred A. Marcus is a professor in the Strategic Management and Organization Department of Carlson School of Management at the University of Minnesota. His recent research projects involve understanding the business implications of environmental challenges and improving environmental decisionmaking. In the past, he has worked on projects about enhancing safety in high-hazard technologies and promoting energy conservation and renewable energy.

In addition to working with companies and governments in the United States and abroad, Marcus has been chair of the Strategic Management and Organization Department at the University of Minnesota; director of the Strategic Management Research Center; a visiting professor at MIT's Sloan School of Management and the Norwegian School of Management; and a research consultant at National Academy of Sciences. He is the author or editor of *Better Environmental Decisions* (with Ken Sexton); *Business and Society: Strategy, Ethics, and the Global Economy*; *Controversial Issues in Energy Policy*; *Managing Environmental Issues* (with Rogene Buchholz and James Post); *Business Strategy and Public Policy* (with Allen M. Kaufman and David R. Beam); and *The Adversary Economy*.

Donald A. Geffen has been a research associate at the Strategic Management Research Center of Carlson School of Management at the University of Minnesota and an independent consultant since 1992. He has worked with Alfred Marcus on projects exploring new competencies for environmental management and regulation and studying pollution prevention responses to the performance-based standards set for sulfur dioxide emissions by the Clean Air Act. From 1982 to 1991, Geffen was a vice president at Alliance Capital Management, an international investment management firm. For more than twenty years before that, he was a professor of physics at the University of Minnesota in Minneapolis, where he participated in interdisciplinary projects regarding the interface among science, technology, and public policy.

Geffen has been a National Science Foundation Postdoctoral Fellow at the Institute for Theoretical Physics in Copenhagen and the European Center for Nuclear Physics (CERN) in Geneva; a research associate at the Carnegie Institute of Technology in Pittsburgh and Oak Ridge National Laboratory in Tennessee; and a visiting professor at the Laboratoire de Physique Théorique et Particules Élémentaires of Université Paris-Sud in Orsay, France, and at Massachusetts Institute of Technology.

Ken Sexton is the Bond Professor of Environmental Health Policy and director of the Center for Environment and Health Policy in the School of Public Health at the University of Minnesota. Before that, he was director of the Office of Health Research at the U.S. Environmental Protection Agency (EPA), where he managed EPA's research programs in toxicology, epidemiology, and exposure monitoring. His research interests focus on risk assessment and risk management, public policies aimed at protecting environmental health, and the role of science in regulatory decisions.

Sexton is the senior editor of two books, *Better Environmental Decisions: Strategies for Governments, Businesses, and Communities* and *Environmental Policy in Transition: Making the Right Choices.*